Handbook of Plant and Crop Physiology

Handbook of Plant and Crop Physiology

Editor

Sunil Kumar

scitus
academics

Handbook of Plant and Crop Physiology

Edited by **Sunil Kumar**

ISBN: 978-1-68117-027-5

Library of Congress Control Number: 2015931824

SCITUS Academics LLC,
www.scitusacademics.com
Box No. 4766, 616 Corporate Way,
Suite 2, Valley Cottage,
NY 10989

Printed in United States of America on Acid Free Paper ∞

Contents

Preface

Plant physiology is the study of the function, or physiology of plants. Fundamental processes such as photosynthesis, respiration, transpiration, and floral induction are studied in plant physiology.

Crop physiology is the study of the plant processes responsible for the growth, development, and production of economic yield by crop plants. Crop physiologists focus on whole plants and plant communities - not individual plant parts, organs, or cells because most of the processes that control yield operate at the whole plant - plant community level.

This book focuses on cellular and molecular aspects of plant and crop physiology, plant and crop physiological responses to heavy metal concentration and agrichemicals, computer modeling in plant physiology, and more. The book provides them with accurate, modern and streamlined content. It includes even more modern applications and color within key images and illustrations to make the material easier to understand.

Editor

Chapter 1

Characterization of Plant Antioxidative System in Response to Abiotic Stresses: A Focus on Heavy Metal Toxicity

Miguel Mourato, Rafaela Reis and Luisa Louro Martins

UIQA, Instituto Superior de Agronomia, Technical University of Lisbon, Lisbon Portugal

INTRODUCTION

During their life span, plants can be subjected to a number of abiotic stresses, like drought, temperature (both high and low), radiation, salinity, soil pH, heavy metals, lack of essential nutrients, air pollutants, etc. When affected by one, or a combination of abiotic stresses, a response is induced by changes in the plant metabolism, growth and general development. Reactive Oxygen Species (ROS) are a natural consequence of the aerobic metabolism, and plants have mechanisms to deal with them in normal conditions, controlling the formation and removal rates. Under stress conditions, cell homeostasis is disrupted and ROS production can increase a lot putting a heavy burden on the those antioxidative mechanisms, some of which are activated in order to eliminate the excess ROS (Mittler et al., 2004). Trace element contamination cause abiotic stress in plants and it can affect crop production and quality. Certain metals, like copper, are essential for plants, but at high concentrations (depending on plant species) can be considered toxic. Other elements like cadmium and arsenic (a metalloid), while not essential elements for plants, are widespread pollutants that are present in nature due to both natural and manmade activities.

Plants have developed different strategies to cope with these stresses. Some use an avoidance strategy to reduce trace element assimilation while others use internal defence mechanisms to cope with the increasing levels of the toxic species. Phytotoxic amounts of trace elements are known to affect several physiological processes and can cause oxidative stress. Plants have developed several trace element defence mechanisms, that allow them to grow despite the presence of variable concentrations of trace elements, but the threshold concentrations as well as the different response mechanisms strongly depend on plant species and on the type of metal. Metal toxicity can cause a redox imbalance and induce the increase of ROS concentration, activating the antioxidant defence mechanisms of plants (Sharma & Dietz, 2009). These mechanisms are very dependent on the metal and the plant but usually include the involvement of the ascorbate-glutathione cycle enzymes which is a major antioxidative defence mechanism, and of other antioxidant enzymes like catalase, peroxidases, and superoxide dismutase. Other non-enzymatic substances with reported antioxidant properties can also be involved in plant defence mechanisms, like ascorbate, glutathione, alkaloids, phenolic compounds, non-proteic amino-acids and carotenoids.

OXIDATIVE STRESS AND ROS PRODUCTION

ROS are produced by all aerobic organisms and are usually kept in balance by the antioxidative mechanisms that exist in all living beings. Because ROS have an important signalling role in plants (Foyer & Noctor, 2003; Vranova et al., 2002), their concentration must be carefully controlled through adequate pathways (Mittler, 2002). ROS can be formed during normal aerobic metabolic processes like photosynthesis and respiration and thus, the majority of ROS are produced in the mitochondria, chloroplast, peroxisomes, plasma membrane and apoplast (Ahmad et al., 2008; Moller, 2001). Other sources of ROS production are NADPH oxidases, amine oxidases and cell-wall peroxidases (Mittler, 2002).

Under certain stress conditions (like excess light, cold, heat, drought, heavy metals etc.) the production of ROS can exceed the capacity of the plant's defence mechanisms, an imbalance in intracellular ROS content is established and this results in oxidative stress (Gill & Tuteja, 2010). Thus, oxidative stress can be defined as the physiological changes resulting from the formation of excess quantities of reactive oxygen species (ROS) (Vangronsveld & Clijsters, 1994). This increase in ROS levels induces a metabolic response in the plant in order to eliminate them. This metabolic response is highly dependent on the plant species, plant growing state and the type and duration of the stress.

Heavy metals[1] are natural elements that are present at different concentrations throughout nature, but whose levels can increase and overtake the toxicity threshold of living beings due to both natural and anthropogenic causes (Sánchez, 2008). As plants must adapt (or die) to the conditions where they grow, the presence of heavy metals can induce oxidative stress and the activation of several defence factors in the plants (Prasad, 2004). It is important to understand how some plants can cope with high concentration of metals in order to produce crops able to grow on contaminated soils (Schröder et al., 2008), to help in environmental cleanup via phytoremediation (Adriano et al., 2004) and to breed plants with higher contents of essential nutrients (Zhao & McGrath, 2009).

Types of ROS

Molecular oxygen (O_2) is in itself a bi-radical[2], as it has two unpaired electrons that have parallel spins (Halliwell, 2006). The ground state of the oxygen molecule is the triplet oxygen ($3O_2$ or ×O-O×), because this is an energetically more favourable state. Due to this electron configuration, it doesn't react easily with organic molecules that have paired electrons with opposite spins. Thus, in order for oxygen to react, it must be activated (Garg & Manchanda, 2009). If the oxygen molecule in its ground state absorbs sufficient energy, the spin of one of the unpaired electrons can be reversed forming singlet oxygen, that can readily react with organic molecules. This can happen during photosynthesis when an excess of light energy cannot be readily dissipated by the photosynthetic machinery (Foyer et al., 1994).

Another form of activation is by partial reduction adding one, two or three electrons giving rise to the superoxide radical, hydrogen peroxide and hydroxyl radical, respectively (Mittler, 2002). The complete reduction of oxygen (adding four electrons) results in water, which is the normal reduction of oxygen that occurs in the mitochondrial electron transport chain, catalyzed by cytochrome oxidase. As such, this type of activation can occur in metabolic pathways that involve an electron transport chain and can thus occur in several cell locations (Alscher et al., 2002).

In Table 1 we present the most important types of ROS. They can be free radicals or nonradicals.

---

[1]Although the term "heavy metal" can be misleading because it includes a whole range of substances some of which do not conform to the more usual chemical definitions of the term, we use it in this text in its environmental context, where it includes metals that are a cause of pollution concern. Other authors sometimes use the expression "trace elements" in this context, although this definition also has problems on its own.

[2]Radicals are usually highly unstable, and thus very reactive, species that have one or more unpaired electrons.

Table 1: Most important types of ROS

Name	Structure	Type	Relative Reactivity
Singlet oxygen	$^1O_2$ (O-O:)	Radical	High
Superoxide	$O_2^{\bullet -}$ (O-O:)	Radical	Medium
Hydrogen peroxide	H_2O_2 (H:O-O:H)	Non-radical	Low
Hydroxyl radical	HO$^{\ddot{Y}}$ (H:O×)	Radical	Very high

Singlet oxygen is mainly produced in the chloroplasts at photosystem II (Asada, 2006) but may also result from lipoxygenase activity and is a highly reactive species that can last for nearly 4 μs in water (Foyer et al., 1994). $^1O_2$ reactivity has as preferred target the conjugated double bonds present on polyunsaturated fatty acids (PUFAs) leaving a specific footprint in the cell (Moller et al. 2007) that can be followed by the detection of several aldehydes like malondialdehyde (MDA) formed by PUFA peroxidation.

The superoxide radical is mainly produced both in the chloroplasts (photosystems I and II) and mitochondria as sub products and in peroxisomes (del Rio et al., 2006; Moller et al., 2007; Rhoads et al., 2006), has a half-life of 2-4 μs and cannot cross phospholipid membranes (Garg & Manchanda, 2009) and so it is important that the cell has adequate in situ mechanism to scavenge this ROS. Superoxide dismutase can catalyse the conversion of this species into hydrogen peroxide. Superoxide radical can also be produced by NADPH oxidase in the plasma membrane (Moller et al. 2007).

Hydrogen peroxide is mainly produced in peroxisomes (del Rio et al., 2006) and also in mitochondria (Rhoads et al., 2006), and also results from the dismutation of superoxide. It is not a radical and can easily cross membranes diffusing across the cell and has a half-life of around 1 ms (Garg & Manchanda, 2009).

The hydroxyl radical, the most reactive of the species listed in Table 1, can be formed from hydrogen peroxide via Fenton and Fenton-like reactions (catalyzed by iron or other transition metals) and, unlike the previous two ROS mentioned, there are no known enzymatic systems able to degrade it (Freinbichler et al., 2011).

Although the superoxide radical and hydrogen peroxide are not as reactive as other species they are produced in large amounts in the cell and can initiate other reactions that lead to more dangerous species (Noctor & Foyer, 1998). In fact, superoxide radical can be converted by specific enzymes into hydrogen peroxide, and this can also be a problem as it cause the occurrence of Fenton reactions (Moller et al., 2007).

Heavy metals are known to induce oxidative stress increasing the ROS concentration. As an example, in figure 1 we present experimental results of work performed by the authors, of the effect of Cd and Cu in hydrogen peroxide content in roots of tobacco plants. As can be seen, there is a good correlation between Cu levels and H_2O_2 content, but the effect of Cd showed only a small non-significant increase. These metals, an essential and a non-essential, do seem to provoke different responses in the plant. The increase in hydrogen peroxide levels with metals is a frequently reported stress indicator (Khatun et al., 2008; Mobin & Khan, 2007).

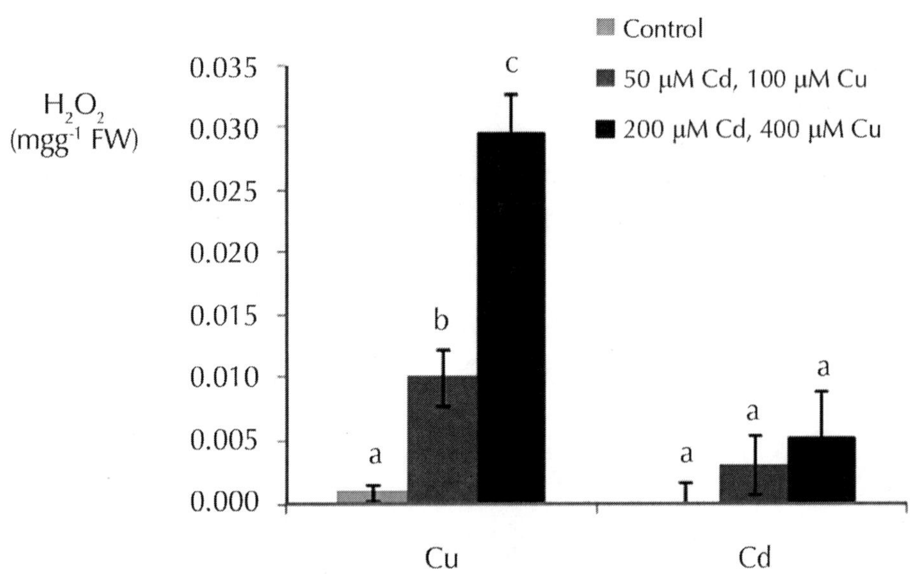

Figure 1: Hydrogen peroxide content in *Nicotiana tabacum* L. plants grown in nutrient solution, with excess copper and cadmium.

ROS Effect in Different Cellular Components (Lipids, DNA, Proteins, and Carbohydrates)

When cell homeostasis is affected by a given stress, ROS production increases to the point where it can damage cellular components and ultimately lead to cell death. ROS can affect lipids, proteins, carbohydrates and DNA and the detailed mechanisms are well detailed in Moller et al. (2007).

Unsaturated fatty acids from lipid membranes are particularly susceptible to ROS oxidation, increasing membrane leakage. Lipid peroxidation occurs through a series of chain reactions that start when a ROS like the hydroxyl radical removes one hydrogen from a carbon from the fatty acid molecule (mainly at the unsaturation). An oxygen can then easily bond to that location forming a lipid peroxyl radical, that can continue and propagate the same kind of reactions (Gill & Tuteja, 2010).

Proteins can also suffer oxidation by ROS, causing certain enzymes to lose its catalytic function. One of the more susceptible targets in proteins are thiol groups the oxidation of which can lead to protein denaturation and loss of functional conformation (Moller et al., 2007). Also, protein oxidation leads to the production of carbonyl groups and to increased rate of proteolysis as the damaged proteins are targeted by proteolytic enzymes (Palma et al., 2002).

Changes in protein content and in protein profile can be found as a consequence of the stress induced by toxic metals. In figure 2, we show the effect of excess copper (50 μM) in the protein content of *Lupinus luteus* leaves, evidencing a significantly higher protein content after 11 days of excess copper, compared to control. In this work, lupin plants were grown in nutrient solution with the indicated Cu concentration. This protein increase could represent the positive balance from the inactivation of some proteins whereas other proteins are formed in relation to the defense response.

In figure 3, we present the protein profile of *Lupinus luteus* leaves after 11 days of exposition to different Cu concentrations. The protein profile showed some changes that can be related to the Cu concentration in nutrient solution. In fact, we found that some protein bands showed higher intensities (56.5 and 17.7 Da) while new protein bands were detected that were not present in control samples (28.5 and 14 Da). These new proteins could be related to the Cu defence mechanism of these plants.

DNA can also be attacked by ROS damaging nucleotide bases, causing mutations and genetic defects (Tuteja et al., 2001).

Both free carbohydrates and wall polysaccharides can react easily with the hydroxyl radical, and this can also be a defence mechanism if the radical reacts with these carbohydrates before damaging more biologically important molecules (Moller et al., 2007).

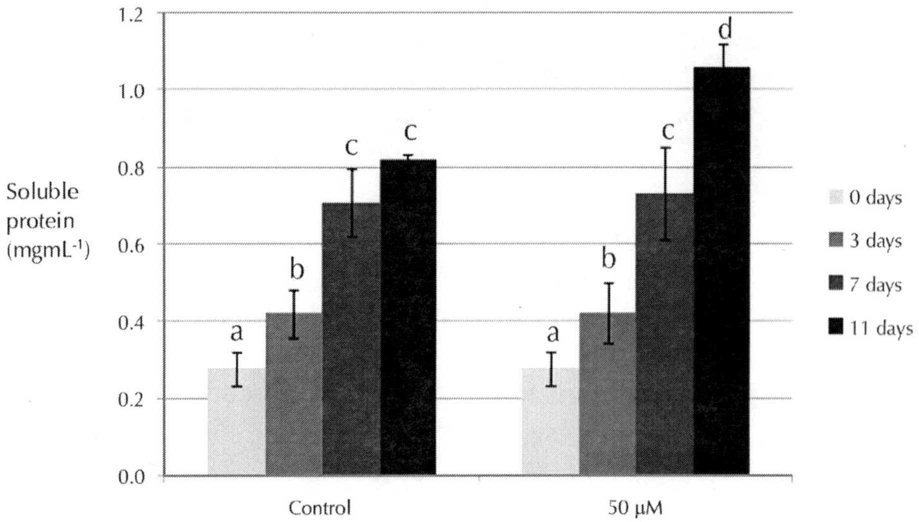

Figure 2: Protein content of *Lupinus luteus* leaves grown in nutrient solution with 0.1 µM of Cu (control) and excess copper (50 µM) for up to 11 days.

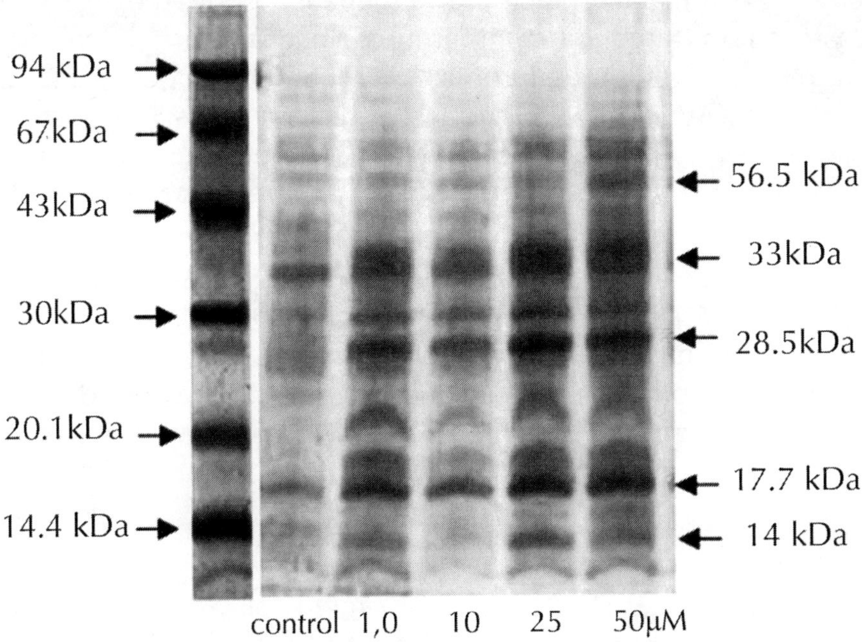

Figure 3: SDS-PAGE protein profile of *Lupinus luteus* leaves after 11 days at different concentrations of copper. Electrophoresis was performed in a 12 % polyacrylamide gel stained with Commassie blue.

ANTIOXIDANT DEFENCE MECHANISMS

Enzymatic Mechanisms

As was said before, enzymatic mechanisms and enzymes involved in specific metabolic pathways are one of the major antioxidative defence strategy of plant defence against excess ROS. Superoxide dismutase (SOD, EC 1.15.1.1), catalyses the dismutation of superoxide molecules into hydrogen peroxide and oxygen (Alscher et al., 2002).

$$O_2^{\bullet-} + O_2^{\bullet-} + 2H^+ \Rightarrow H_2O_2 + O_2$$

SOD has a metal cofactor and depending on the metal can be classified in three different groups, localized in different cell compartments: FeSOD (chloroplasts), MnSOD (mitochondria and peroxisomes), Cu/ZnSOD (chloroplast and cytosol). As SOD produces hydrogen peroxide that is subsequently converted to water by peroxidases and catalases, the activity of all these enzymes must be carefully balanced.

Catalase (CAT, EC 1.11.1.6) exists mainly in the peroxisomes and as during stress the number of these organelles increase, CAT can have an important role in H_2O_2 detoxification that can diffuse into the peroxisome from other cell locations where it is produced (Mittler, 2002). CAT catalyses the hydrogen peroxide breakdown to water:

$$H_2O_2 + H_2O_2 \Rightarrow 2H_2O + O_2$$

Peroxidases (EC 1.11.1) are a member of a large family of enzymes that are ubiquitous in the cell and have numerous roles in plant metabolism (Passardi et al., 2005), namely to remove hydrogen peroxide formed due to induced stress using different reductants. They have the general reaction:

$$H_2O_2 + R(OH)_2 \Rightarrow 2H_2O + RO_2$$

$R(OH)_2$ represents different electron donors: guaiacol peroxidase (GPOD, EC 1.11.1.7) uses mainly phenolic donors, ascorbate peroxidase (APX, EC 1.11.1.11) uses ascorbic acid and glutathione peroxidase (GPX, EC 1.11.1.9) uses glutathione.

Besides their role as a scavenger of hydrogen peroxide, cell-wall peroxidases are also involved in ROS formation, both as a defence against biotic stresses and as a signaling process against several stresses, leading to the activation of other defence mechanisms (Mika et al., 2004).

APX has a much higher affinity to H_2O_2 than CAT suggesting that they have different roles in the scavenging of this ROS, with APX being responsible for maintaining the low levels of hydrogen peroxide while CAT is responsible for the removal of its excess (Mittler, 2002). The water-water cycle (Figure 4) occurs in chloroplasts and is a fundamental mechanism to avoid photooxidative damage (Rizhsky et al., 2003), using SOD and APX to scavenge the superoxide radical and hydrogen peroxide in the location where they are produced avoiding the deleterious effects of their reactivity with other cellular components (Asada, 1999; Shigeoka et al., 2002).

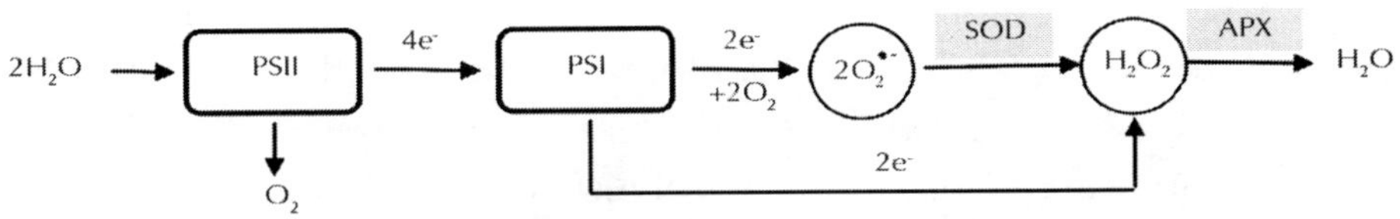

Figure 4: The water-water cycle. PSI and PSII - Photosystems I and II, SOD – Superoxide dismutase, APX - ascorbate peroxidase.

The ascorbate-glutathione cycle (Figure 5) is an important group of reactions involved in ROS detoxification, as it converts hydrogen peroxide (formed as a consequence of an induced stress or via SOD action) and occurs in several cell compartments, like chloroplasts, cytosol, mitochondria, peroxisomes and apoplast. It uses APX and GPX as well as other enzymes like monodehydroascorbate reductase (MDHAR, EC 1.6.5.4) and dehydroascorbate reductase (DHAR, EC 1.8.5.1) that have a role in the regeneration of the reduced form of ascorbate and glutathione reductase (GR, EC 1.6.4.2), important to maintain the pool of reduced glutathione.

Other enzymes like heme oxygenase (HO, EC 1.14.99.3), that catalyzes the stereo specific cleavage of heme to biliverdin (Balestrasse et al., 2005) have been reported to have a role in plant defence mechanisms against oxidative stress.

Enzymatic Responses to Heavy Metal Stress

There has been extensive studies on the activities of the enzymes involved in ROS defence on plants subjected to heavy metal stress (Sharma & Dietz, 2009). Although different metals can cause oxidative stress, their mode of action is different. For example while copper, an essential element toxic at high concentrations, is involved in redox reactions, cadmium, which has no known biological function, cannot. However, they both can induce oxidative stress, through different mechanisms

(Cuypers et al., 2010). Of all the main enzymes involved in oxidative stress defence, like SOD, APX, CAT and peroxidases, published reports both describe an increase in its activity and a decrease (or no change), depending on plant species, plant organ, type of metal, duration of the treatment, plant age, and growing media (Gratão et al., 2005).

In table 2 we list some representative publications of this kind of studies, regarding determinations made at a single time of growth (time series studies show also changes along the time, further complicating the analysis). As can be seen, there is a huge variation between the behaviour of the enzymes involved in oxidative stress. In some situations the activity increases for lower concentrations of the metal and then decreases as the defence mechanism breaks down due to excessive concentration. This also shows that different components of the antioxidative defence system described above are active under different conditions. The increase in activity in a given enzyme can be a signal that that enzyme has been activated or its expression upregulated. On the other hand sometimes the effect of the metal can be so drastic that enzymes structures are being affected with a consequent decrease in activity. Several enzymes have metal cofactors so there could be a link between these enzymes expression and metal availability (Cohu & Pilon, 2007).

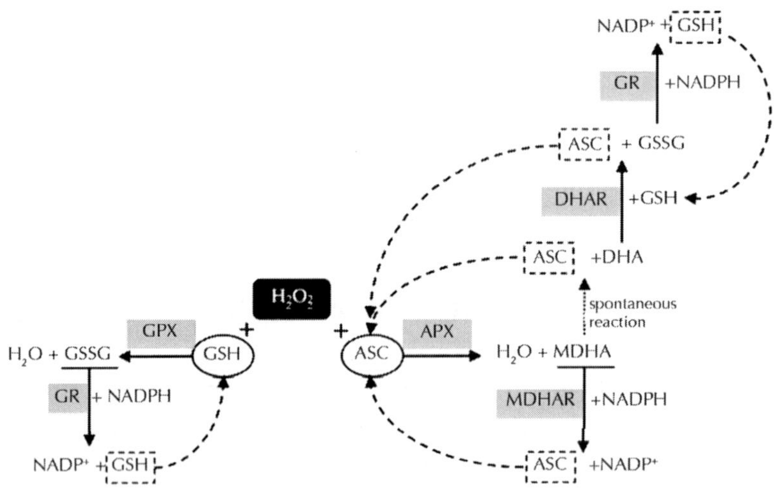

Figure 5: The ascorbate-glutathione cycle. Non-enzymatic compounds: ASC - ascorbate, MDHA - monodehydroascorbate, DHA - dehydroascorbate, GSH

- glutathione (reduced), GSSG - glutathione (oxidized). Enzymes (grey box): APX - ascorbate peroxidase, GPX -glutathione peroxidase, GR - glutathione reductase, MDHAR – monodehydroascorbate reductase, DHAR - dehydroascorbate reductase.

Consequently, enzymatic response can be complex to analyse. For example, in figure 6A we present the activity of guaiacol peroxidase in tomato roots growing for 3 days with 50 μM Cu in nutrient solution. As can be seen by the relative activity, no significant changes in POD activity were detected compared to control. However, the isoperoxidase profile (figure 6B) showed the appearance of new isoforms in tomato roots (C, D), while other isoforms showed less intensity (A, B). These results indicate that although enzymatic activity can be similar in control and stressed plants it is possible that some isoforms can be activated in response to excess copper.

Polyphenol oxidase (PPO, EC 1.10.3.1) is an oxidoreductase that catalyzes the oxidation of phenols to quinones and its activity has been shown to increase under heavy metal stress and has thus been associated to some form of defence mechanism (Ali et al., 2006; L. L. Martins & Mourato, 2006). Kováčik et al. (2009) observed an increase in root PPO activity with Cu and Cd and concluded that the formation of polymerized phenols could be used to complex free metal ions. On the other hand PPO has also been associated to a catalase-like activity (Gerdemann et al., 2001), and could thus have a role in direct hydrogen peroxide removal.

Peroxiredoxins (PRX, EC 1.11.1.15) are ubiquitous antioxidant enzymes that participate in cellular redox homeostasis and reduce hydrogen peroxide to water. PRX levels have been shown to increase under several abiotic stresses suggesting a role in the defence mechanisms (Barranco-Medina et al., 2007).

Table 2: Changes in enzyme activities for several plants and metals. The symbols after the enzymes indicate if its activity increased (), decreased (¯), remained the same (=) or increased for lower concentrations and decreased for higher concentrations or vice versa (¯ or ¯)

Plant Species	Metal	Concentration	Organ	Enzyme	References
Arabidopsis thaliana	Cd	5, 10, 20 μM	leaves	APX ¯, GPOD ¯, SOD =, CAT =, GR ¯	(Smeets et al., 2008)
Brassica juncea	Cd	10, 25, 50 μM	leaves	APX ¯, GPOD ¯, CAT , GR =, APX ¯, CAT ¯,	(Nouairi et al., 2009)
''	Cd	10, 30, 50, 100 μM	leaves	GR ¯, GPX ¯, MDHAR ¯, DHAR ¯	(Markovska et al., 2009)
''	As	5, 25 μM	leaves	APX , SOD , CAT =	(Khan et al., 2009)
''	Cd	5, 15, 35 $mg.kg^{-1}$	leaves	GPOD , APX ¯, CAT ¯	(Pinto et al., 2009)
Brassica napus	Cd	10, 25, 50 μM	leaves	APX ¯, GPOD , CAT ¯, GR ¯	(Nouairi et al., 2009)
Cannabis sativa	Cd	25, 50, 100 $mg.kg^{-1}$	seedlings	GPOD , SOD , CAT =	(Shi et al., 2009)
Daucus carota	Cu	100, 200, 400, 800 $mg.kg^{-1}$	leaves	APX ¯, SOD ¯, CAT ¯	(Ke et al., 2007)
Elsholtzia splendens	Cu	25, 50, 100, 500 μM	leaves	APX ¯, GPOD ¯, SOD ¯, CAT ¯	(Peng et al., 2006)
Matricaria chamomilla	Cd	3, 60, 120 μM	Leaves, roots	CAT , GR	(Kovácik & Backor, 2008)
''	Cu	3, 60, 120 μM	Leaves, roots	CAT , GR	(Kovácik & Backor, 2008)
Nicotiana tabacum	Cd	10, 25, 50, 100 μM	Young leaves old leaves	GPOD , SOD ¯ GPOD , SOD ¯	(Martins et al., 2011)

,,	Cd	5, 15, 35 mg.kg^{-1}	leaves	GPOD =, APX $^{-}$, CAT	(Pinto et al., 2009)
Solanum nigrum	Cd	5, 15, 35 mg.kg^{-1}	leaves	GPOD , APX , CAT	(Pinto et al., 2009)
Typha angustifolia	Cd	1 mM	leaves	GPOD , APX =, GPX =, SOD , CAT =	(Bah et al., 2011)
,,	Cr	2 mM	leaves	GPOD , APX =, GPX =, SOD , CAT =	(Bah et al., 2011)
,,	Pb	1 mM	leaves	GPOD , APX =, GPX =, SOD , CAT =	(Bah et al., 2011)
Withania somnifera	Cu	10, 25, 50, 100, 200 µM	leaves	APX $^{-}$, GPOD , SOD $^{-}$, CAT^{-}, GR $^{-}$, MDHAR , DHAR $^{-}$	(Khatun et al., 2008)
Zea mays	Cd	300, 600, 900 µM	leaves	APX $^{-}$, GPOD $^{-}$, SOD , GR $^{-}$	(Ekmekci et al., 2008)

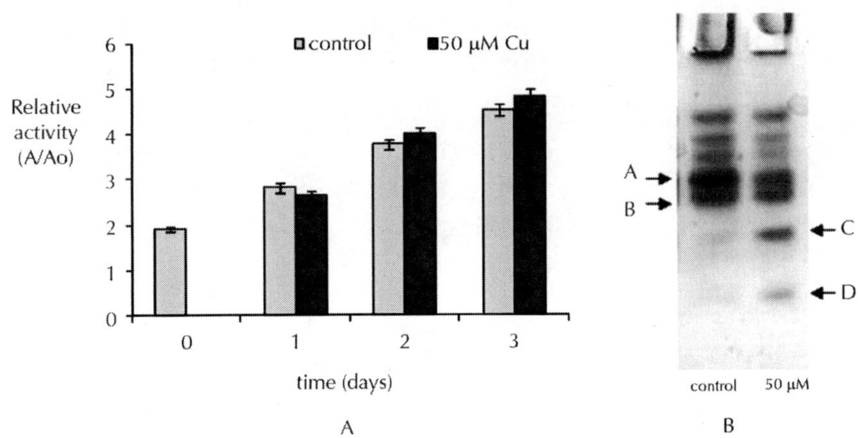

Figure 6: Guaiacol peroxidase activity of tomato roots for 3 days of 50 µM Cu in nutrient solution (A) and the correspondent isoperoxidase profile for plants collected after 3 days (B). Cu concentration in control plants was 0.1 µM.

Non Enzymatic Mechanisms

Besides the enzymatic mechanisms described in the two previous chapters, there is a whole range of other substances that have been reported to be involved in antioxidative defence in plants. Some are well known and have been extensively studied (like the role of ascorbate and glutathione) while others are thought to be part of defence mechanisms but its role remains to be fully understood. While many studies report the increase in the concentration of some substance in relation to the induction of a stress, this type of correlation is not, by itself, conclusive enough to ascertain the effect of that substance in the plant metabolism. Of course, when the levels of the compounds described in this chapter increase, most of the enzymes involved in the respective biosynthesis are also induced.

Several organic molecules have been reported to be able to form complexes with heavy metals, like phytochelatins (see section 3.4), organic acids and amino acids like proline. But exactly what kind of molecules complex the metals seem to be highly dependent not only on metal type and plant species but also on plant organ and compartment, as different complexes can be formed in the cytoplasm, the xylem and phloem, for example (Sharma & Dietz, 2006).

Ascorbate and glutathione are antioxidants that exist in relatively high concentration in cell compartments (Potters et al., 2002), and are involved in the ascorbate-glutathione cycle as described above being also a substrate for APX an enzyme important in H_2O_2 removal.

Ascorbate (ASC) is an electron donor that can be oxidized to the radical monodehydroascorbate (MDHA) and this compound can then form dehydroascorbate (DHA) (Figure 7):

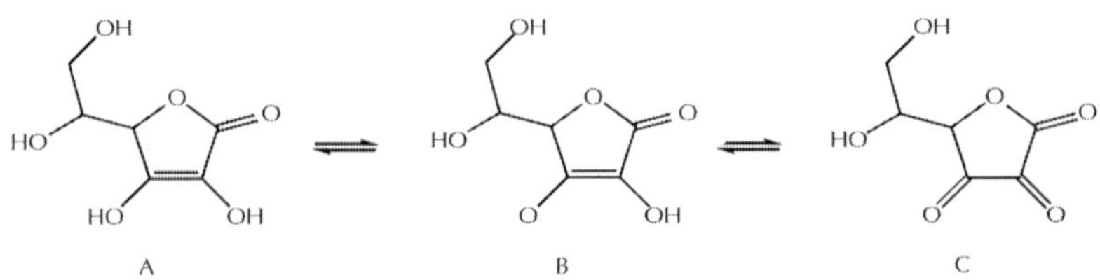

Figure 7: Structure of ascorbate (A), monodehydroascorbate (B) and dehydroascorbate (C).

Besides its role as an enzyme substrate, ascorbate also reacts directly with singlet oxygen and superoxide and is important in the regeneration of a-tocopherol and certain carotenoids (Potters et al., 2002).

Glutathione (Figure 8) is a tripeptide (containing glutamate, cysteine and glycine) that can exist in two predominant forms: the reduced form (usually represented by GSH) and the oxidized form (usually represented by GSSG) (Noctor & Foyer, 1998). It is involved in the sulphur metabolism and in defence reactions against oxidative stress (Potters et al., 2002). It can also lead to the synthesis of phytochelatins that are important sequesters for certain heavy metals (Cobbett & Goldsbrough, 2002).

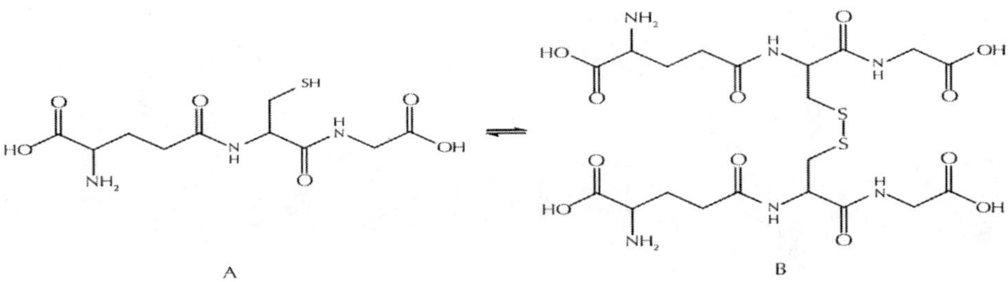

Figure 8: Structure of glutathione in its reduced (A) and oxidized forms (B)

Heat shock proteins (HSP) are proteins that not only showed increased expression under heat stress but are also molecular chaperones that protect other proteins from stress induced damage (Feder & Hofmann, 1999). HSP help proteins maintain or recover their native conformation, and remove potentially harmful polypeptides. HSP expression has been associated to high temperature, cold, drought, light, heavy metals, salt and ozone stresses (Timperio et al., 2008). Kochhar & Kochhar (2005) detected the induction of both high and low molecular weight HSPs in response to combined heat and cadmium stress. ROS also have a signaling role during a stress, in order to induce HSP production (Timperio et al., 2008).

Carotenoids have an important protective role during photosynthesis as these molecules can quench the excited states of chlorophyll in order to avoid the production of singlet oxygen. As a consequence, the carotenoid molecules become themselves excited but this is not a big problem as they don't have enough energy to form this ROS species (Taiz & Zeiger, 2002). Terpenoids are a large class of organic compounds derived from the isoprene unit that could also have an antioxidative role in plants, although that is not yet clear (Grassmann et al., 2002). Flavonoids are organic molecules with a structure similar to flavone (Figure 9), that have been shown to have a protective role against several stresses (Jaakola et al., 2004), both by themselves and in conjugation with peroxidases (Mika et al., 2004). Anthocyanins, a type of flavonoids (they are glucosides of anthocyanidins, Figure 9) present in the vacuoles, have an antioxidative capacity (Kahkonen & Heinonen, 2003) but its location in the cell prevents them to contact directly with ROS production sites, although its levels have been reported to increase under Cd stress (Mobin & Khan, 2007).

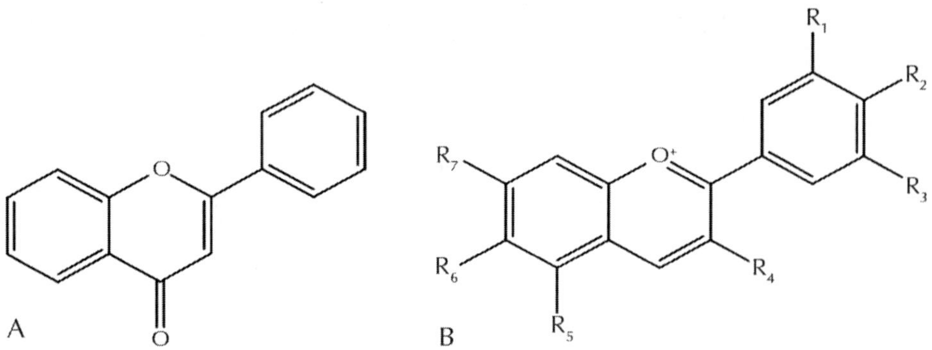

Figure 9: Structure of flavone (A) and of anthocyanidins (B, where R_1 toR_7 can be H, OH or OCH_3 according to the exact type of anthocyanidin).

Thiols can play an important antioxidative role, protecting membrane lipids. Lipoic acid (Figure 10), both in its reduced and oxidized form, is reported to have antioxidative properties due to its direct scavenging of ROS. It is also able to chelate several metal ions that induces oxidative stress (Navari-Izzo et al., 2002) and thus can have an important role in cell protection.

Tocopherols are a class of compounds synthesized by photosynthetic organisms that have vitamin E activity. a-Tocopherol (Figure 11) is the most common form in leaves while g-tocopherol is more common in roots (Abbasi et al., 2007). They have a role in ROS protection as they can quench singlet oxygen (Gill & Tuteja, 2010) and can act as an antioxidant and terminate chain reactions occurring during lipid peroxidation (Garg & Manchanda, 2009).

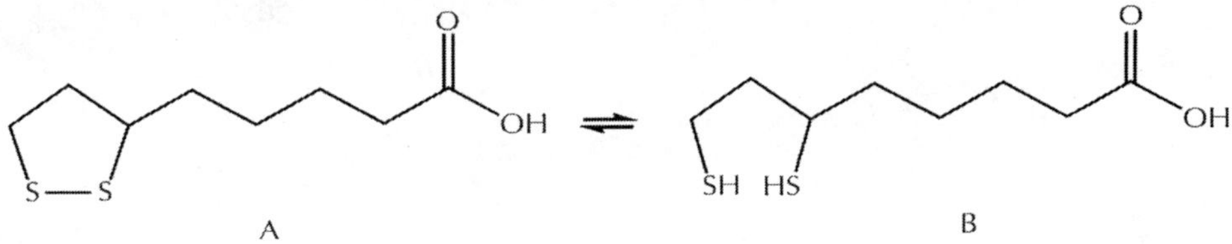

Figure 10: Structure of lipoic acid in its oxidized (A) and reduced (B) forms.

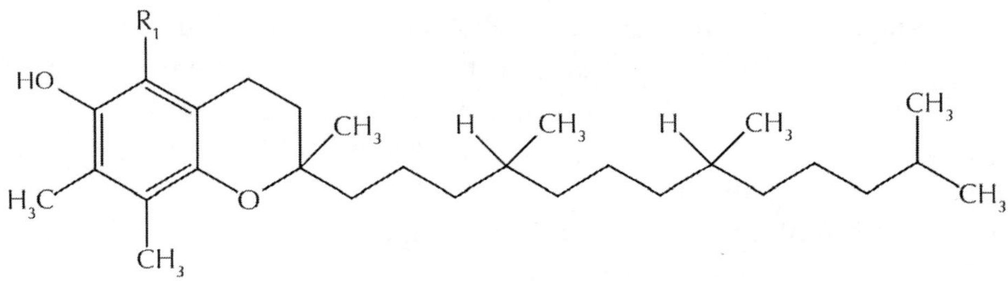

Figure 11: Common structure of a-tocopherol (with R_1=CH_3) and of g-tocopherol (with R_1 = H).

Proline (an amino acid, shown in figure 12A), is a compatible solute that participate in the osmotic adjustment of plant cells being able to balance water stress. Proline has been reported to improve plant resistance to oxidative stress by scavenging ROS (namely by quenching singlet oxygen and hydroxyl radicals), increasing the activity of antioxidative enzymes and protecting them and maintaining redox homeostasis (Matysik et al., 2002), and they could also participate in signalling pathways that regulate stress related genes (Khedr et al., 2003). Although the concentration of proline has been shown to increase under heavy metal stress in certain plants (Martins et al., 2011), its exact role in heavy metal detoxification is unclear as, under certain conditions, it could be only an indirect response due to heavy metal induced disturbances in plant water balance (Schat et al., 1997).

Histidine (figure 12B) is another amino acid that has been mainly linked to Ni hyper accumulator plants (Sharma & Dietz, 2006).

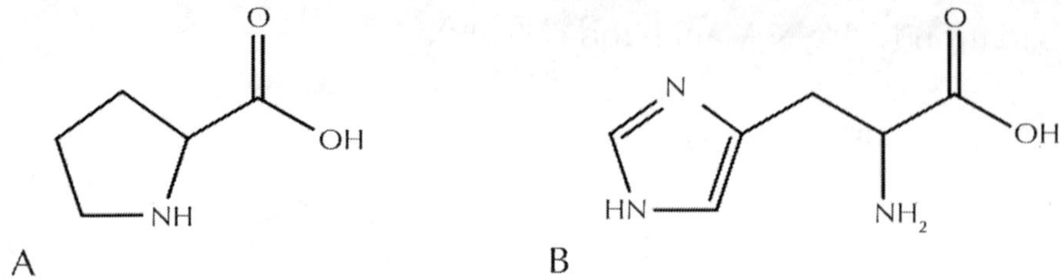

Figure 12: Structure of proline (A) and histidine (B).

Glycine betaine (2-trimethylammonioacetate, figure 13A) is another compatible solute that is involved in plant resistance against abiotic stresses, namely salt stress (Banu et al., 2009). It helps not only in controlling water balance but can also help to maintain protein and membrane structure (F.-L. Zhang et al., 2008). Nicotianamine (figure 13B) is an amino acid that has a known role in heavy metal transport in plants (Stephan & Scholz, 1993), but recent findings have suggested an important role also in heavy metal tolerance namely in relation to hyper accumulating species (Sharma & Dietz, 2006). Mugineic acid (figure 13C) is a siderophore (that is, an iron-chelating compound) that promotes iron acquisition from the rhizosphere, but that may also participate in the distribution of other metals in the plant (Haydon & Cobbett, 2007). Meda et al. (2007) reports that phytosiderophores can alleviate Cd toxicity in maize.

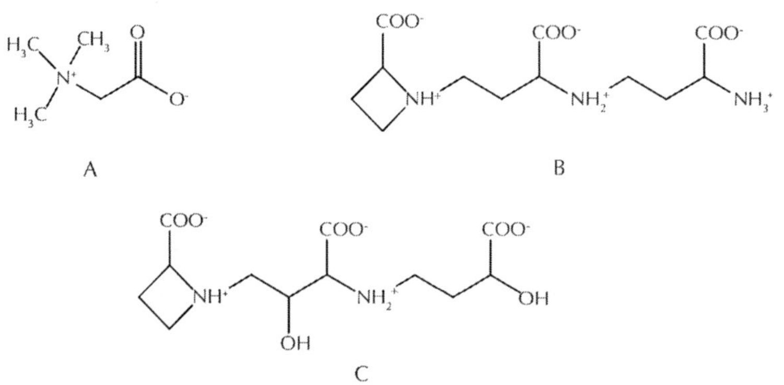

Figure 13: Structure of glycine betaine (A), nicotianamine (B) and mugineic acid (C).

Polyamines are low molecular weight amines (figure 14) that have a role in plant growth and developmental processes (Kakkar & Sawhney, 2002). The concentration of some polyamines has been reported to increase under abiotic stress. Mascher (2002) showed that the concentration of putrescine, spermidine and spermine incresaed when red clover plants were subjected to As toxicity. However it is still not clear the exact role these compounds play in heavy metal defence, but a participation in the stabilization and protection of the membrane systems has been proposed (Sharma & Dietz, 2006).

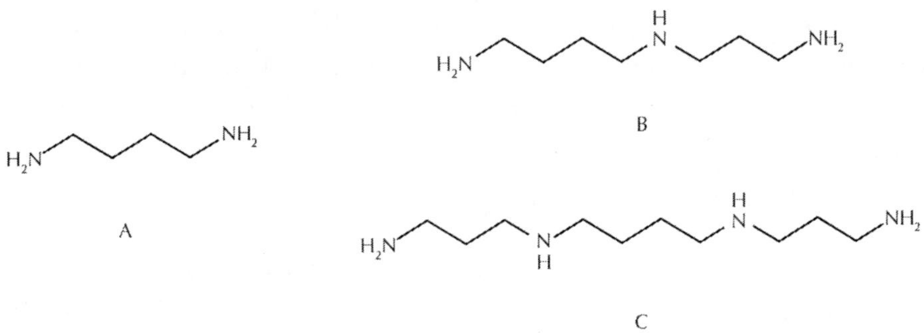

Figure 14: Structure of several polyamines. A - putrescine (a di-amine); B - spermidine (a triamine); C - spermine (a tetra-amine).

Although soluble sugars have been linked to metabolic pathways that produce ROS, they can also have an important role in ROS scavenging mechanisms. Increased glucose levels can increase the production of NADPH (via the pentose-phosphate pathway), that is an important intermediate in the ascorbate-glutathione cycle (Couée et al., 2006) as NADPH is the primary electron donor that assures a intracellular reduction status. Both glucose and sucrose levels have been shown to increase in some plant species under certain abiotic stresses but it is not obvious that this happens due to a putative defence mechanism induction (Couée et al., 2006), although these sugars also participate in signalling mechanisms. Van den Ende and Valluru (2009) suggest that sucrose might have a protective role against stress due to its capacity to scavenge ROS. Other sugars like raffinose (Figure 15A) and fructans (which are fructose polymers) are also reported to have a protective role of membranes against several stresses, namely freezing and drought stress (Van den Ende & Valluru, 2009).

Trehalose (Figure 15B) is a non-reducing disaccharide that can also participate in the stabilization of membranes and protection of proteins under abiotic stresses (Luo et al., 2008). Although trehalose exists in numerous organisms (like bacteria, fungi and nematodes) it is not found widespread in plants, and when it is found is usually in very low concentrations (Wingler, 2002), but several studies have correlated the availability of this disaccharide (or the expression of the genes related to its synthesis) with stress responses.

Almeida et al. (2005) found that transgenic tobacco plants over-expressing trehalose-6-phosphate synthase had higher resistance against different abiotic stresses (like drought and temperature). Nery et al. (2008) stated that trehalose participated in protecting cells against hydrogen peroxide, by preventing oxidation of both membranes and proteins.

Brassinosteroids are a family of polyhydroxysteroids that have been reported to modify the activity of antioxidant enzymes and the level of non-enzymatic compounds (like ascorbic acid and tocopherols), when plants are subjected to different abiotic stresses, but its effect is still poorly understood (Bajguz & Hayat, 2009). ROS also have a signalling role in hormone responses, like ABA, auxin and ethylene (Kwak et al., 2006).

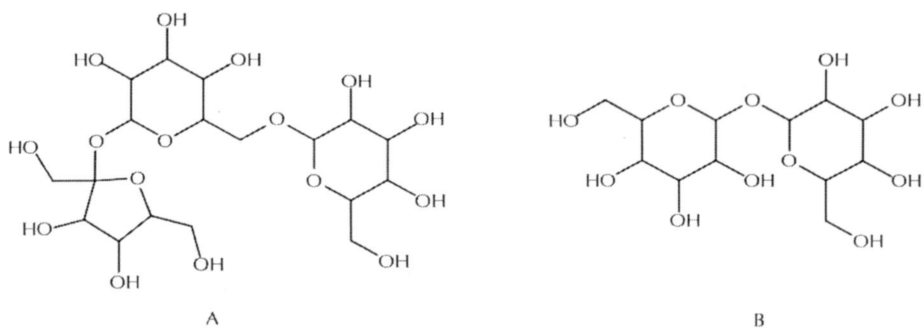

Figure 15: Structure of raffinose (A) and trehalose (B).

Other Non Enzymatic Substances Involved in Heavy Metal Tolerance

Several other substances, not described above, have been reported to be involved in the tolerance mechanisms against heavy metal tolerance. Metals are rarely available in the free ionic form in plants but are bound to different types of organic molecules. These include organic acids (like citric and maleic acid) and some amino acids (Haydon & Cobbett, 2007).

In fact, organic acids, among other organic solutes of several types, can accumulate in leaves in stress conditions, such as water stress, with an important role on subcellular structures protection acting as osmolytes (Pinheiro et al., 2004). One possible effect of heavy metal stress is that plants restrict water uptake, thus indirectly inducing water stress.

Any changes in the concentrations of the intermediate organic acids can reflect an influence of a heavy metal in metabolic pathways as they are involved in primary plant metabolism such as cell respiration and the formation of ATP. Organic acids can have a detoxification role complexing metals, and reducing their availability to the plant, but this role is not yet fully understood (Hall, 2002).

The metabolism of several phenols has been reported to change under Cu and Cd stress indicating a putative role of these compounds in heavy metal detoxification by decreasing the presence of the free metal ions (Kováčik et al., 2009). Lignin accumulation has also been described as a consequence of Cu toxicity (Lequeux et al., 2010) but whether this is related to a heavy metal defence mechanism is still not clear.

Zorrig et al. (2010) suggested a transport role for citrate to translocate Cd from the roots to the shoots of lettuce plants while Panfili et al. (2009) also studied the effect of citrate in the uptake of Cd. Citrate and malate have been reported to be major ligands for Zn and Ni in several studies (Haydon & Cobbett, 2007).

Phytochelatins (PC) are small metal-binding peptides with the structure (γ-Glu-Cys)n-Gly (figure 16) where n ranges between 2 and 11 (Cobbett & Goldsbrough, 2002). Its synthesis is catalyzed by phytochelatin synthase using glutathione as a substrate (Clemens, 2006;

Grill et al., 1989). They form complexes with metals like Cd (due to the high affinity metals have to the thiol group present in cysteine) and sequester them to the vacuoles and are thus an important mechanism that plants use to avoid heavy metal toxicity. Other peptides with a structure similar to phytochelatins have been identified, with the terminal Gly being replaced by other amino acids like serine, glutamic acid and glutamine in the case of isophytochelatins or alanine in homophytochelatins (Oven et al., 2002). All of these peptides have been reported to participate in the detoxification mechanisms of various plant species against several metals or metalloids, besides Cd, like As (Vázquez et al., 2009) and Pb (Z. C. Zhang et al., 2008).

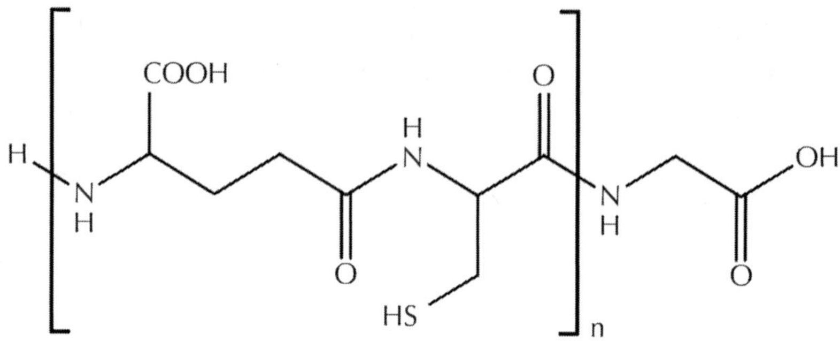

Figure 16: General structure of phytochelatins (n ranges between 2 and 11).

Metallothioneins (MT) are other small peptides and in plants they contain typically between 60 to 85 amino acids (Freisinger, 2009), also containing cysteine in various amounts according to the type of MT. They have a different synthesis pathway from phytochelatins (Cobbett & Goldsbrough, 2002), and are thought to be important in metal complexation, but can also have other roles like ROS scavenging (Hassinen et al., 2011). MTs are reported to be much more important in Cu tolerance of certain plants than PCs (Mijovilovich et al., 2009).

ACKNOWLEDGMENTS

The authors acknowledge the financial support of Fundação para a Ciência e Tecnologia, through its project PTDC/AGR-AAM/102821/2008 and the Research Unit Environmental Chemistry.

REFERENCES

1. Abbasi, A.-R., Hajirezaei, M., Hofius, D., Sonnewald, U., & Voll, L. M. (2007). Specific Roles of {alpha}- and {gamma}-Tocopherol in Abiotic Stress Responses of Transgenic Tobacco. Plant Physiology, Vol.143, No.4, 1720-1738.
2. Adriano, D. C., Wenzel, W. W., Vangronsveld, J., & Bolan, N. S. (2004). Role of assisted natural remediation in environmental cleanup. Geoderma, Vol.122, 121-142.
3. Ahmad, P., Sarwat, M., & Sharma, S. (2008). Reactive oxygen species, antioxidants and signaling in plants. Journal of Plant Biology, Vol.51, No.3, 167-173.
4. Almeida, A. M., Villalobos, E., Araújo, S. S., Leyman, B. V. D. P., Alfaro-Cardoso, L., Fevereiro, P. S., Torné, J. M., & Santos, D. M. (2005). Transformation of tobacco with an *Arabidopsis thaliana* gene involved in trehalose biosynthesis increases tolerance to several abiotic stresses. Euphytica, Vol.146, 165-176.
5. Alscher, R. G., Erturk, N., & Heath, L. S. (2002). Role of superoxide dismutases (SODs) in controlling oxidative stress in plants. Journal of Experimental Botany, Vol.53, No.372, 1331-1341.
6. Asada, K. (1999). The Water-Water Cycle in Chloroplasts: Scavenging of Active Oxygens and Dissipation of Excess Photons. Annual Review of Plant Physiology & Plant Molecular Biology, Vol.50, No.1, 601.
7. Asada, K. (2006). Production and scavenging of reactive oxygen species in chloroplasts and their functions. Plant Physiology, Vol.141, No.2, 391-396.
8. Bah, A. M., Dai, H., Zhao, J., Sun, H., Cao, F., Zhang, G., & Wu, F. (2011). Effects of Cadmium, Chromium and Lead on Growth, Metal Uptake and Antioxidative Capacity in *Typha angustifolia*. Biological Trace Element Research, Vol.142, No.1, 77-92.
9. Bajguz, A., & Hayat, S. (2009). Effects of brassinosteroids on the plant responses to environmental stresses. Plant Physiology and Biochemistry, Vol.47, No.1, 1-8.
10. Balestrasse, K. B., Noriega, G. O., Batlle, A., & Tomaro, M. L. (2005). Involvement of heme oxygenase as antioxidant defense

in soybean nodules. Free Radical Research, Vol.39, No.2, 145-151.

11. Banu, M. N. A., Hoque, M. A., Watanabe-Sugimoto, M., Matsuoka, K., Nakamura, Y., Shimoishi, Y., & Murata, Y. (2009). Proline and glycinebetaine induce antioxidant defense gene expression and suppress cell death in cultured tobacco cells under salt stress. Journal of Plant Physiology, Vol.166, No.2, 146-156.
12. Barranco-Medina, S., Krell, T., Finkemeier, I., Sevilla, F., Lazaro, J.-J., & Dietz, K.-J. (2007). Biochemical and molecular characterization of the mitochondrial peroxiredoxin PsPrxII F from Pisum sativum. Plant Physiology and Biochemistry, Vol.45, No.10-11, 729-739.
13. Clemens, S. (2006). Evolution and function of phytochelatin synthases. Journal of Plant Physiology, Vol.163, No.3, 319-332.
14. Cobbett, C., & Goldsbrough, P. (2002). Phytochelatins and metallothioneins: Roles in heavy metal detoxification and homeostasis. Annual Review of Plant Biology, Vol.53, 159-182.
15. Cohu, C. M., & Pilon, M. (2007). Regulation of superoxide dismutase expression by copper availability. Physiologia Plantarum, Vol.129, No.4, 747-755.
16. Couée, I., Sulmon, C., Gouesbet, G., & El Amrani, A. (2006). Involvement of soluble sugars in reactive oxygen species balance and responses to oxidative stress in plants. Journal of Experimental Botany, Vol.57, No.3, 449-459.
17. Cuypers, A., Plusquin, M., Remans, T., Jozefczak, M., Keunen, E., Gielen, H., Opdenakker, K., Nair, A., Munters, E., Artois, T., Nawrot, T., Vangronsveld, J., & Smeets, K. (2010). Cadmium stress: an oxidative challenge. BioMetals, Vol.23, No.5, 927-940.
18. del Rio, L. A., Sandalio, L. M., Corpas, F. J., Palma, J. M., & Barroso, J. B. (2006). Reactive oxygen species and reactive nitrogen species in peroxisomes. Production, scavenging, and role in cell signaling. Plant Physiology, Vol.141, No.2, 330-335.
19. Ekmekci, Y., Tanyolac, D., & Ayhan, B. (2008). Effects of cadmium on antioxidant enzyme and photosynthetic activities in leaves of two maize cultivars. Journal of Plant Physiology, Vol.165, No.6, 600-611.

20. Feder, M. E., & Hofmann, G. E. (1999). Heat-shock proteins, molecular chaperones, and the stress response: Evolutionary and ecological physiology. Annual Review of Physiology, Vol.61, 243-282.

21. Foyer, C. H., Lelandais, M., & Kunert, K. J. (1994). Photooxidative Stress in Plants. Physiologia Plantarum, Vol.92, No.4, 696-717.

22. Foyer, C. H., & Noctor, G. (2003). Redox sensing and signalling associated with reactive oxygen in chloroplasts, peroxisomes and mitochondria. Physiologia Plantarum, Vol.119, No.3, 355-364.

23. Freinbichler, W., Colivicchi, M. A., Stefanini, C., Bianchi, L., Ballini, C., Misini, B., Weinberger, P., Linert, W., Vareslija, D., Tipton, K. F., & Della Corte, L. (2011). Highly reactive oxygen species: detection, formation, and possible functions. Cellular and Molecular Life Sciences, Vol.68, No.12, 2067-2079.

24. Freisinger, E. (2009). Metallothioneins in Plants. In A. Sigel, H. Sigel & R. K. O. Sigel (Eds.), Metallothioneins and Related Chelators (Vol. 5, pp. 107-153). Cambridge: The Royal Society of Chemistry.

25. Garg, N., & Manchanda, G. (2009). ROS generation in plants: Boon or bane? Plant Biosystems, Vol.143, No.1, 81 - 96.

26. Gill, S. S., & Tuteja, N. (2010). Reactive oxygen species and antioxidant machinery in abiotic stress tolerance in crop plants. Plant Physiology and Biochemistry, Vol.48, No.12, 909-930.

27. Grassmann, J., Hippeli, S., & Elstner, E. F. (2002). Plant's defence and its benefits for animals and medicine: role of phenolics and terpenoids in avoiding oxygen stress. Plant Physiology and Biochemistry, Vol.40, No.6-8, 471-478.

28. Gratão, P. L., Polle, A., Lea, P. J., & Azevedo, R. A. (2005). Making the life of heavy metal stressed plants a little easier. Functional Plant Biology, Vol.32, 481-494.

29. Grill, E., Loffler, S., Winnacker, E. L., & Zenk, M. H. (1989). Phytochelatins, the Heavy-MetalBinding Peptides of Plants, Are Synthesized from Glutathione by a Specific Gamma Glutamylcysteine Dipeptidyl Transpeptidase (Phytochelatin Synthase). Proceedings of the National Academy of Sciences of the United States of America, Vol.86, No.18, 6838-6842.

30. Hall, J. L. (2002). Cellular mechanisms for heavy metal detoxification and tolerance. Journal of Experimental Botany, Vol.53, No.366, 1-11.
31. Halliwell, B. (2006). Reactive species and antioxidants. Redox biology is a fundamental theme of aerobic life. Plant Physiology, Vol.141, No.2, 312-322.
32. Hassinen, V. H., Tervahauta, A. I., Schat, H., & Karenlampi, S. O. (2011). Plant metallothioneins - metal chelators with ROS scavenging activity? Plant Biology, Vol.13, No.2, 225-232.
33. Haydon, M. J., & Cobbett, C. S. (2007). Transporters of ligands for essential metal ions in plants. New Phytologist, Vol.174, No.3, 499-506.
34. Jaakola, L., Määttä-Riihinen, K., Kärenlampi, S., & Hohtola, A. (2004). Activation of flavonoid biosynthesis by solar radiation in bilberry (Vaccinium myrtillus L.) leaves. Planta, Vol.218, No.5, 721-728.
35. Kahkonen, M. P., & Heinonen, M. (2003). Antioxidant activity of anthocyanins and their aglycons. Journal of Agricultural and Food Chemistry, Vol.51, No.3, 628-633.
36. Kakkar, R. K., & Sawhney, V. K. (2002). Polyamine research in plants - a changing perspective. Physiologia Plantarum, Vol.116, No.3, 281-292.
37. Ke, W. S., Xiong, Z. T., Xie, M. J., & Luo, Q. (2007). Accumulation, subcellular localization and ecophysiological responses to copper stress in two *Daucus carota* L. populations. Plant and Soil, Vol.292, No.1-2, 291-304.
38. Khan, I., Ahmad, A., & Iqbal, M. (2009). Modulation of antioxidant defence system for arsenic detoxification in Indian mustard. Ecotoxicology and Environmental Safety, Vol.72, No.2, 626-634.
39. Khatun, S., Ali, M. B., Hahn, E.-J., & Paek, K.-Y. (2008). Copper toxicity in *Withania somnifera*: Growth and antioxidant enzymes responses of in vitro grown plants. Environmental and Experimental Botany, Vol.64, No.3, 279-285.
40. Khedr, A. H. A., Abbas, M. A., Wahid, A. A. A., Quick, W. P., & Abogadallah, G. M. (2003). Proline induces the expression of salt stress responsive proteins and may improve the adaptation of

Pancratium maritimum L. to salt stress. Journal of Experimental Botany, Vol.54, No.392, 2553-2562.

41. Kochhar, S., & Kochhar, V. K. (2005). Expression of antioxidant enzymes and heat shock proteins in relation to combined stress of cadmium and heat in Vigna mungo seedlings. Plant Science, Vol.168, No.4, 921-929.

42. Kovácik, J., & Backor, M. (2008). Oxidative status of *Matricaria chamomilla* plants related to cadmium and copper uptake. Ecotoxicology, Vol.17, No.6, 471-479.

43. Kováčik, J., Klejdus, B., Hedbavny, J., Štork, F., & Bačkor, M. (2009). Comparison of cadmium and copper effect on phenolic metabolism, mineral nutrients and stress-related parameters in *Matricaria chamomilla* plants. Plant and Soil, Vol.320, No.1, 231-242.

44. Kwak, J. M., Nguyen, V., & Schroeder, J. I. (2006). The role of reactive oxygen species in hormonal responses. Plant Physiology, Vol.141, No.2, 323-329.

45. Lequeux, H., Hermans, C., Lutts, S., & Verbruggen, N. (2010). Response to copper excess in *Arabidopsis thaliana*: Impact on the root system architecture, hormone distribution, lignin accumulation and mineral profile. Plant Physiology and Biochemistry, Vol.48, No.8, 673-682.

46. Luo, Y., Li, W.-M., & Wang, W. (2008). Trehalose: Protector of antioxidant enzymes or reactive oxygen species scavenger under heat stress? Environmental and Experimental Botany, Vol.63, No.1-3, 378-384.

47. Markovska, Y. K., Gorinova, N. I., Nedkovska, M. P., & Miteva, K. M. (2009). Cadmiuminduced oxidative damage and antioxidant responses in *Brassica juncea* plants. Biologia Plantarum, Vol.53, No.1, 151-154.

48. Martins, L. L., Mourato, M. P., Cardoso, A. I., Pinto, A. P., Mota, A. M., Goncalves, M. d. L. S., & de Varennes, A. (2011). Oxidative stress induced by cadmium in *Nicotiana tabacum* L.: effects on growth parameters, oxidative damage and antioxidant responses in different plant parts. Acta Physiologiae Plantarum, Vol.33, No.4, 1375-1383.

49. Mascher, R., Lippmann, B., Holzinger, S., & Bergmann, H. (2002). Arsenate toxicity: effects on oxidative stress response molecules and enzymes in red clover plants. Plant Science, Vol.163, No.5, 961-969.

50. Matysik, J., Alia, Bhalu, B., & Mohanty, P. (2002). Molecular mechanisms of quenching of reactive oxygen species by proline under stress in plants. Current Science, Vol.82, No.5, 525-532.

51. Meda, A. R., Scheuermann, E. B., Prechsl, U. E., Erenoglu, B., Schaaf, G., Hayen, H., Weber, G., & von Wiren, N. (2007). Iron Acquisition by Phytosiderophores Contributes to Cadmium Tolerance. Plant Physiology, Vol.143, No.4, 1761-1773.

52. Mijovilovich, A., Leitenmaier, B., Meyer-Klaucke, W., Kroneck, P. M. H., Gotz, B., & Kupper, H. (2009). Complexation and Toxicity of Copper in Higher Plants. II. Different Mechanisms for Copper versus Cadmium Detoxification in the Copper-Sensitive Cadmium/Zinc Hyperaccumulator Thlaspi caerulescens (Ganges Ecotype). Plant Physiology, Vol.151, No.2, 715-731.

53. Mika, A., Minibayeva, F., Beckett, R., & Lüthje, S. (2004). Possible functions of extracellular peroxidases in stress-induced generation and detoxification of active oxygen species. Phytochemistry Reviews, Vol.3, No.1, 173-193.

54. Mittler, R. (2002). Oxidative stress, antioxidants and stress tolerance. Trends in Plant Science, Vol.7, No.9, 405-410.

55. Mittler, R., Vanderauwera, S., Gollery, M., & Van Breusegem, F. (2004). Reactive oxygen gene network of plants. Trends in Plant Science, Vol.9, No.10, 490-498.

56. Mobin, M., & Khan, N. A. (2007). Photosynthetic activity, pigment composition and antioxidative response of two mustard (*Brassica juncea*) cultivars differing in photosynthetic capacity subjected to cadmium stress. Journal of Plant Physiology, Vol.164, No.5, 601-610.

57. Moller, I. M. (2001). Plant mitochondria and oxidative stress: Electron transport, NADPH turnover, and metabolism of reactive oxygen species. Annual Review of Plant Physiology and Plant Molecular Biology, Vol.52, 561-591.

58. Moller, I. M., Jensen, P. E., & Hansson, A. (2007). Oxidative modifications to cellular components in plants. Annual Review of Plant Biology, Vol.58, 459-481.

59. Navari-Izzo, F., Quartacci, M. F., & Sgherri, C. (2002). Lipoic acid: a unique antioxidant in the detoxification of activated oxygen species. Plant Physiology and Biochemistry, Vol.40, No.6-8, 463-470.
60. Nery, D. d. C. M., da Silva, C. G., Mariani, D., Fernandes, P. N., Pereira, M. D., Panek, A. D., & Eleutherio, E. C. A. (2008). The role of trehalose and its transporter in protection against reactive oxygen species. Biochim Biophys Acta, Vol.1780, No.12, 1408-1411.
61. Noctor, G., & Foyer, C. H. (1998). Ascorbate and glutathione: Keeping active oxygen under control. Annual Review of Plant Physiology and Plant Molecular Biology, Vol.49, 249-279.
62. Nouairi, I., Ben Ammar, W., Ben Youssef, N., Ben Miled, D. D., Ghorbal, M., & Zarrouk, M. (2009). Antioxidant defense system in leaves of Indian mustard (*Brassica juncea*) and rape (*Brassica napus*) under cadmium stress. Acta Physiologiae Plantarum, Vol.31, No.2, 237-247.
63. Oven, M., Page, J. E., Zenk, M. H., & Kutchan, T. M. (2002). Molecular characterization of the homo-phytochelatin synthase of soybean Glycine max - Relation to phytochelatin synthase. Journal of Biological Chemistry, Vol.277, No.7, 4747-4754.
64. Palma, J. M., Sandalio, L. M., Corpas, F. J., Romero-Puertas, M. C., McCarthy, I., & del Rio, L. A. (2002). Plant proteases, protein degradation, and oxidative stress: role of peroxisomes. Plant Physiology and Biochemistry, Vol.40, No.6-8, 521-530.
65. Panfili, F., Schneider, A., Vives, A., Perrot, F., Hubert, P., & Pellerin, S. (2009). Cadmium uptake by durum wheat in presence of citrate. Plant and Soil, Vol.316, No.1, 299-309.
66. Passardi, F., Cosio, C., Penel, C., & Dunand, C. (2005). Peroxidases have more functions than a Swiss army knife. Plant Cell Reports, Vol.24, No.5, 255-265.
67. Peng, H. Y., Yang, X. E., Yang, M. J., & Tian, S. K. (2006). Responses of antioxidant enzyme system to copper toxicity and copper detoxification in the leaves of Elsholtzia splenden. Journal of Plant Nutrition, Vol.29, No.9, 1619-1635.
68. Pinheiro, C., Passarinho, J. A., & Ricardo, C. P. (2004). Effect of drought and rewatering on the metabolism of Lupinus albus organs. Journal of Plant Physiology, Vol.161, No.11, 1203-1210.

69. Pinto, A. P., Alves, A. S., Candeias, A. J., Cardoso, A. I., de Varennes, A., Martins, L. L., Mourato, M. P., Goncalves, M. L. S., & Mota, A. M. (2009). Cadmium accumulation and antioxidative defences in *Brassica juncea* L. Czern, *Nicotiana tabacum* L. and *Solanum nigrum* L. International Journal of Environmental Analytical Chemistry, Vol.89, No.8-12, 661-676.
70. Potters, G., De Gara, L., Asard, H., & Horemans, N. (2002). Ascorbate and glutathione: guardians of the cell cycle, partners in crime? Plant Physiology and Biochemistry, Vol.40, No.6-8, 537-548. Prasad, M. N. V. (2004). Heavy Metal Stress in Plants: Springer.
71. Rhoads, D. M., Umbach, A. L., Subbaiah, C. C., & Siedow, J. N. (2006). Mitochondrial reactive oxygen species. Contribution to oxidative stress and interorganellar signaling. Plant Physiology, Vol.141, No.2, 357-366.
72. Rizhsky, L., Liang, H. J., & Mittler, R. (2003). The water-water cycle is essential for chloroplast protection in the absence of stress. Journal of Biological Chemistry, Vol.278, No.40, 38921-38925.
73. Sánchez, M. L. (2008). Causes And Effects Of Heavy Metal Pollution. New York: Nova Science Publishers, Inc.
74. Schat, H., Sharma, S. S., & Vooijs, R. (1997). Heavy metal-induced accumulation of free proline in a metal-tolerant and a nontolerant ecotype of Silene vulgaris. Physiologia Plantarum, Vol.101, No.3, 477-482.
75. Schröder, P., Herzig, R., Bojinov, B., Ruttens, A., Nehnevajova, E., Stamatiadis, S., Memon, A., Vassilev, A., Caviezel, M., & Vangronsveld, J. (2008). Bioenergy to save the world. Environmental Science and Pollution Research, Vol.15, No.3, 196-204.
76. Sharma, S. S., & Dietz, K.-J. (2009). The relationship between metal toxicity and cellular redox imbalance. Trends in Plant Science, Vol.14, No.1, 43-50.
77. Sharma, S. S., & Dietz, K. J. (2006). The significance of amino acids and amino acid-derived molecules in plant responses and adaptation to heavy metal stress. Journal of Experimental Botany, Vol.57, No.4, 711-726.

78. Shi, G. R., Cai, Q. S., Liu, Q. Q., & Wu, L. (2009). Salicylic acid-mediated alleviation of cadmium toxicity in hemp plants in relation to cadmium uptake, photosynthesis, and antioxidant enzymes. Acta Physiologiae Plantarum, Vol.31, No.5, 969-977.
79. Shigeoka, S., Ishikawa, T., Tamoi, M., Miyagawa, Y., Takeda, T., Yabuta, Y., & Yoshimura, K. (2002). Regulation and function of ascorbate peroxidase isoenzymes. Journal of Experimental Botany, Vol.53, No.372, 1305-1319.
80. Smeets, K., Ruytinx, J., Semane, B., Van Belleghem, F., Remans, T., Van Sanden, S., Vangronsveld, J., & Cuypers, A. (2008). Cadmium-induced transcriptional and enzymatic alterations related to oxidative stress. Environmental and Experimental Botany, Vol.63, No.1-3, 1-8.
81. Stephan, U. W., & Scholz, G. (1993). Nicotianamine - Mediator of Transport of Iron and Heavy-Metals in the Phloem. Physiologia Plantarum, Vol.88, No.3, 522-529.
82. Taiz, L., & Zeiger, E. (2002). Plant Physiology (3rd ed.). Sunderland, MA: Sinuaer Associates, Inc.
83. Timperio, A. M., Egidi, M. G., & Zolla, L. (2008). Proteomics applied on plant abiotic stresses: Role of heat shock proteins (HSP). Journal of Proteomics, Vol.71, No.4, 391-411.
84. Tuteja, N., Singh, M. B., Misra, M. K., Bhalla, P. L., & Tuteja, R. (2001). Molecular mechanisms of DNA damage and repair: Progress in plants. Critical Reviews in Biochemistry and Molecular Biology, Vol.36, No.4, 337-397.
85. Van den Ende, W., & Valluru, R. (2009). Sucrose, sucrosyl oligosaccharides, and oxidative stress: scavenging and salvaging? Journal of Experimental Botany, Vol.60, No.1, 9-18.
86. Vangronsveld, J., & Clijsters, H. (1994). Toxic effects of metals. In M. E. Farago (Ed.), Plants and the chemical elements. Biochemistry, uptake, tolerance and toxicity (pp. 149-177). Weinheim: VCH Verlagsgesellschaft.
87. Vázquez, S., Goldsbrough, P., & Carpena, R. O. (2009). Comparative analysis of the contribution of phytochelatins to cadmium and arsenic tolerance in soybean and white lupin. Plant Physiology and Biochemistry, Vol.47, No.1, 63-67.

88. Vranova, E., Inze, D., & Van Breusegem, F. (2002). Signal transduction during oxidative stress. Journal of Experimental Botany, Vol.53, No.372, 1227-1236.

89. Wingler, A. (2002). The function of trehalose biosynthesis in plants. Phytochemistry, Vol.60, No.5, 437-440.

90. Zhang, F.-L., Niu, B., Wang, Y.-C., Chen, F., Wang, S.-H., Xu, Y., Jiang, L.-D., Gao, S., Wu, J., Tang, L., & Jia, Y.-J. (2008). A novel betaine aldehyde dehydrogenase gene from Jatropha curcas, encoding an enzyme implicated in adaptation to environmental stress. Plant Science, Vol.174, No.5, 510-518.

91. Zhang, Z. C., Gao, X., & Qiu, B. S. (2008). Detection of phytochelatins in the hyperaccumulator Sedum alfredii exposed to cadmium and lead. Phytochemistry, Vol.69, No.4, 911-918.

92. Zhao, F.-J., & McGrath, S. P. (2009). Biofortification and phytoremediation. Current Opinion in Plant Biology, Vol.12, No.3, 373-380.

93. Zorrig, W., Rouached, A., Shahzad, Z., Abdelly, C., Davidian, J.-C., & Berthomieu, P. (2010). Identification of three relationships linking cadmium accumulation to cadmium tolerance and zinc and citrate accumulation in lettuce. Journal of Plant Physiology, Vol.167, No.15, 1239-1247.

Chapter 2

Genetic and Molecular Aspects of Plant Response to Drought in Annual Crop Species

Anna M. De Leonardis, Maria Petrarulo, Pasquale De Vita, and Anna M. Mastrangelo

CRA-Cereal Research Centre, Foggia, Italy

INTRODUCTION

Stress is defined as any soil and climatic conditions or combination of both that hinders the full realization of genetic potential of a plant, limiting their growth, development and reproduction. These effects in plants of agricultural interest have a major impact on productivity and quality and thus represent, together with biotic stress, the cause of the gap between yield potential and actual production (Ciais et al.,

2005). Stressful environmental conditions are extreme air temperature, drought, excessive presence of salts, anoxia and hypoxia, ozone and heavy metals. Among these factors, heavy damages on agricultural production in Mediterranean environments are exerted by drought, salt stress and early spring low temperatures. The changes in climate forecasted for the near future are expected to exacerbate the onset and magnitude of events of stress due to increased drought and erratic rainfall and rise of evapotranspiration rates due to growing temperatures. Responses to drought are species specific and often genotype specific (De Leonardis et al., 2007). Moreover, the nature of drought response of plants is influenced by the duration and severity of water loss (Pinheiro & Chaves, 2011), the age and stage of development at the point of drought exposure (De Leonardis et al., 2007), as well as the organ and cell type experiencing water deficits (Pastori & Foyer, 2002).

Plants use various mechanisms to cope with drought stress including their morphology, physiology and metabolism at organ and cellular levels (Levitt, 1972). The Figure 1 shows the drought response strategies which include i) escape, ii) avoidance, and iii) tolerance. Escaping strategy, via a short life cycle or developmental plasticity (Araus et al., 2002), allows the plant to complete its life cycle during the period of sufficient water supply before the onset of drought. The drought avoidance mechanism, via enhanced water uptake and reduced water loss (Chaves et al., 2002), involves strategies which help the plant to maintain high water status during periods of stress, either by a more efficient water absorption from roots or by reducing evapotranspiration from aerial parts. Drought tolerance, via osmotic adjustment, enhanced antioxidative capacity and physical desiccation tolerance of the organs, allows to withstand water deficit with low tissue water potential (Ingram & Bartels, 1996). The osmotic compounds synthesized include proteins and aminoacids (like proline, aspartic acid and glutamic acid), methylated quaternary ammonium compounds (e.g. glycine betaine, alanine betaine), hydrophilic proteins (e.g. late embryogenesis abundant (LEA), carbohydrates (like fructan and sucrose) and cyclitols (e.g. D-pinitol, mannitol).

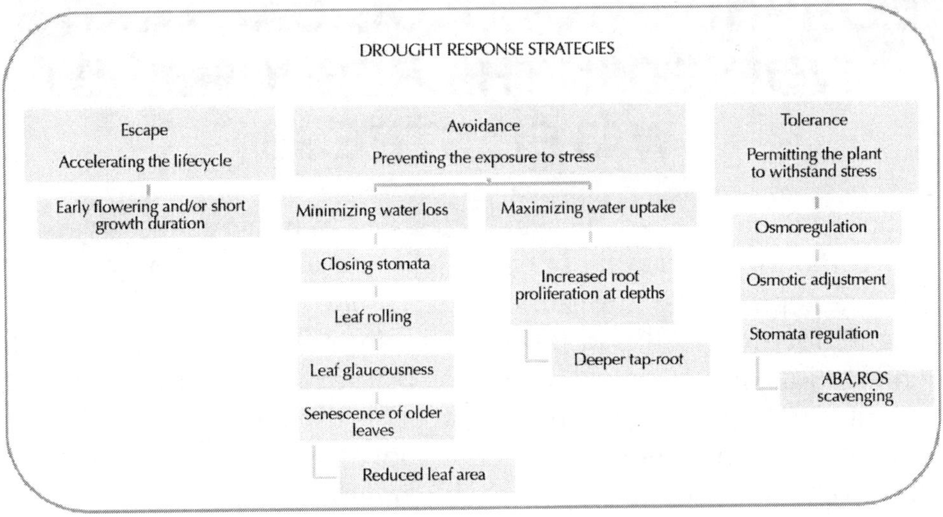

Figure 1: Plant drought response mechanisms and main related traits.

Recent research has uncovered physiological-, biochemical- and molecular-based mechanisms involved in the drought response in plants (Amudha & Balasubramani, 2011). More research into how plants respond to drought conditions is needed and will become more important in the future based on climate change predictions of an increase in arid areas (Petit et al., 1999). Understanding plant responses to drought is of great importance in order to select plants more tolerant to stress (Reddy et al., 2004). Advances in the understanding of these processes may lead to genetically modified drought tolerant crop plants. This chapter focuses on the most recent findings on water stress response in plants. Both morpho-physiological traits and molecular changes contribute to promote stress resistance.

In particular, the future perspectives of breeding for drought tolerance are viewed as resulting from the integration of genomic approaches based on the identification of genomic regions involved in the control of stress-related traits and a deep knowledge of the molecular mechanisms acting at cellular level in response to drought stress.

MORPHO-PHYSIOLOGICAL TRAITS INVOLVED IN THE RESPONSE TO WATER STRESS

As the damage exerted by water stress is translated into important loss in amount and quality of crop yield, the improvement of drought tolerance represented and still represents one of the major objectives of plant breeding. At this purpose, a very important task consists of the identification of the main phenotypic features for plant to cope with drought, and therefore the formulation of the drought-tolerant ideotype. Physiological traits relevant for the responses to water deficits and/or modified by water deficits span a wide range of vital processes.

Morphological traits as early plant vigour, wider leaves and a more prostate growth habit can sustain a rapid ground cover thus avoiding loss of water by soil evaporation and suppressing weed competition for water, with a clear advantage on maintaining a favourable plant water status in order to sustain transpiration and yield (Mastrangelo et al., 2011a).

Plant phenology (escape) represents an important aspect for selecting drought tolerant crops, as it allows the alignment of plant life cycle to the features of the target drought environment. In this regard, the genetic improvement of crops has to take into account the modality of drought stress occurrence in the various environments, and in particular the stress timing, frequency and intensity. As an example, earliness is an effective breeding strategy for enhancing yield stability in Mediterranean environments where crops are exposed to terminal drought stress, even if an extreme earliness leads to yield penalty (Cattivelli et al., 1994). Nevertheless, in the case of cereal species in environments in which the drought stress is experienced in early season during the initial vegetative stage, late flowering, followed by a short grain-filling period, can lead to higher yield (van Ginkel et al., 1998). However, early-flowering varieties will escape terminal drought, but they are not necessarily considered drought-resistant.

One basic mechanism for reducing the impact of drought is early stomatal closure at the beginning of a period of water deficit. Stomatal closure reduces water loss, but also reduces the gas exchange between the plant and the ambient air. The reduced CO_2 intake then results in reduced photosynthesis (Chaves et al., 2002). Nevertheless high yield requires high stomatal conductance to sustain a great CO_2 fixation. Some leaf traits, such as stomatal number/density and leaf mesophyll structure can be important in increasing the water use efficiency. In particular, studies carried out in wheat suggest that the high-yielding modern varieties are "opportunistic", that is they have high rates of stomatal conductance with optimal soil moisture, but markedly reduce stomatal conductance when soil moisture is limiting (Siddique et al., 1990; Rizza et al., 2011). Maximal rates of photosynthesis were also positively correlated with increased yields of advanced varieties, while leaf temperatures were negatively correlated (Fischer et al., 1998). Leaf permeability is another crucial trait, as leaves can lose water through cuticle, increasing crop transpiration without an associated benefit in CO_2 fixation. Glaucousness, which is caused by the presence of epicuticular wax, can prevent these losses (Kerstiens, 2006).

At cellular level, osmotic adjustment is an adaptive mechanism in which the accumulation of solutes helps to maintain a favourable gradient of water potential in the soil-plant-air system. It allows to maintain a sufficient water absorption from a relatively dry soil for sustaining photosynthetic and transpiration activity, and cell expansion for root growth (Mastrangelo et al., 2011a; Dichio et al., 2006). Regarding the importance of this trait in improving grain yield in water stressed environment, a positive correlation between osmotic adjustment and yield increases has been found in particular in conditions of severe water stress (Serraj & Sinclair, 2002).

Besides above-ground traits of plant, deep rooted cultivars have demonstrated a clear yield advantage under water stress conditions. An increased root development in presence of water stress represents a complementary strategy to stomatal closure regulation. The influence of root architecture on yield and other agronomic traits, especially under stress conditions, has been widely reported in all major crops (Tuberosa et al., 2002a; de Dorlodot et al., 2007). A deep and expanded root system should permit to explore a greater soil volume and extract more water. The information available on the genetic control of root traits in the field and their relationships with yield is limited, mainly due to difficulty of measuring root characteristics in a large number of genotypes. Moreover, field studies on roots often require destructive approaches and are complicated by heterogeneity in soil profile, structure, and composition. The acquisition and analysis of root parameters such as total root length are tedious, time consuming, and often inaccurate (Zoon & Van Tienderen, 1990). Furthermore, environmental effects on root development have been documented by a number of researcher. Many of the root characteristics, such as length, average diameter, surface area, and mass have been used to asses the quantity of roots and the functional fraction of the root system. Total root mass is usually viewed as easier to measure than root length or surface area and has frequently been used to compare root systems. However, total root mass alone cannot describe many root functions adequately involved in plant-soil relationship. In the case of annual crops capture of water, at sowing for establishment and late in the season for grain filling, may be the most important target for root system traits. The location and the timing of these water sources within the soil profile depend on the soil type and its water holding capacity, the preceding crop and its water use, the soil water content at sowing and the pattern of rainfall during and after the crop growing season. Modelling can estimate when and where valuable water is likely to be present in the soil profile for targeting root traits (Lilley & Kirkegaard, 2007; Sadras & Rodriguez, 2007). Late-season water, for example, may be located mid-profile, or at the bottom of the root zone. This suggests that it would be beneficial to combine root vigour with other root characteristics to favour resource capture.

The other characteristics may include weak root gravitropism to promote a more widespreading root system for shallower water uptake, or a strong gravitropism to promote deeper root penetration and deep-water uptake (Ho et al., 2005; Manschadi et al., 2006), faster extension towards moisture (hydrotropism) (Eapen et al., 2005) and more or less nodal and seminal root axes (Hochholdinger et al., 2004). Root growth in soil can be limited by physical, chemical, and biological properties of the soil. Despite the intense work carried out on these topics, there is still insufficient understanding upon the soil factors which limit root growth, and the influence of time period and weather conditions on them. Without this information, it is difficult to manage soil to maximize crop production. In terms of physical limitations to root growth, water stress (too little water for root growth), hypoxia or anoxia (too little oxygen), and mechanical impedance (soil that is too hard for roots to penetrate rapidly) are the major causes of poor root system growth and development. In particular, there is a strong interplay between the strength and water content of soil. As soils dry, capillary forces make matric potential more negative, often causing strength to increase rapidly (Whitmore & Whalley, 2009). A review from Bengough et al. (2011) describes selectively both old and new literature on root elongation in drying soil and the role of water stress, mechanical impedance, and their likely interactions.

BREEDING FOR DROUGHT TOLERANCE IMPROVEMENT

Drought tolerance has been historically one of the major targets of genetic improvement of crops, and some relevant results have been obtained during the last century despite the low heritability, due to a high genotype x environment (G x E) interaction, of this trait. Consistent genetic gains (from 10 to 50 kg ha^{-1} yr^{1}) have been registered for cereals and legumes over the last century in all countries, including those characterized by vast drought-prone regions (Calderini & Slafer, 1998; Abeledo et al., 2002). Many studies suggest that cultivars selected for high yield in stress free environments are also adapted to stress prone environments (Cattivelli et al., 2008). In different field experiments modern durum wheat genotypes outperformed the old ones in all test environments including those with moderate drought stress and showed a stronger responsiveness to improved fertility (De Vita et al., 2010). Moderate drought stress is defined physiologically as reduced cell turgor that generally results in reduced stomatal conductance (reduced water loss from the leaf), and lower cellular water potential, which allows the tissue to hold onto the water that is in the leaf more tenaciously (Levitt 1972). This suggests that some of the traits selected to improve potential yield can still sustain yield at least in mild to moderate drought conditions ensuring yield stability (Slafer et al., 2005; Tambussi et al., 2005). A possible explanation is that the main targets of selection (high harvest index in wheat and barley, stay green in maize and sorghum, resistance to pests and diseases, nitrogen use efficiency) are equally useful under dry and wet conditions and, often, the best performances for these traits were overriding the differences in drought adaptability (Mastrangelo et al., 2011a). In some cases adaptive traits were shown to contribute significantly to performance under drought.

Retrospective studies on maize showed that most of the genetic yield improvement for hybrids bred in the second half of the last century could be attributed to traits related to tolerance to stress, like high plant population density, weed interference, low night temperatures during the grain-filling period, low soil moisture, and low soil N (Cattivelli et al., 2008; Tollenaar & Wu, 1999; Tollenaar & Lee, 2002).

Because of the complex nature of drought tolerance, conventional breeding has obtained little success in this regard. Successful cases of genetic improvement for yield in droughtprone environments have been obtained by selecting for secondary traits related to drought tolerance. In maize the silk-tassel interval was identified as a highly indicative secondary trait for drought-resistant breeding (Bolanos & Edmeades, 1996). Spikelet fertility can be visually estimated under field conditions and has been used as an indirect index for drought screening in rice (Garrity & O'Toole, 1994). Another example is based on the use of carbon isotope discrimination (Δ) as a surrogate for water use efficiency to select wheat lines with high water use efficiency in drought-prone environments (Rebetzke et al., 2002). During photosynthesis plants discriminate against the heavy isotope of carbon (^{13}C) and, as a result, in several C3 species, Δ is positively correlated with the ratio of internal leaf CO_2 concentration to ambient CO_2 concentration (Ci/Ca) and negatively associated with transpiration efficiency. Thus, a high Ci/Ca leads to a higher Δ and a lower transpiration efficiency (Farquhar & Richards, 1984). In the last years a great effort has been devoted to the identification of genomic regions involved in the control of traits related to drought stress tolerance. Once the region has been mapped, closely linked molecular markers are identified, which can be used in breeding programs based on Marked Assisted Selection (MAS). The wide range of physiological and biochemical mechanisms involved in dehydration response explains the complexity of plant response to drought, for which a high number of quantitative trait loci (QTLs) widespread on many chromosomes have been found (Cattivelli et al., 2008).

As an example, Yang et al., (2007) reported several QTLs for accumulation and remobilization of water-soluble carbohydrates in wheat stems. Depending on cultivars and environments, stem water-soluble carbohydrates accumulated before flowering, and during the early periods after flowering, contributed up to 70% or more of the grain weight under terminal drought conditions (Yang et al., 2001). Major genomic regions controlling productivity and related traits (Carbon isotope ratio, osmotic potential, chlorophyll content, flag leaf rolling index) were identified on chromosomes 2B, 4A, 5A and 7B by Peleg et al., (2009) in durum wheat. QTLs for productivity were associated with QTLs for droughtadaptive traits, suggesting the involvement of several strategies in wheat adaptation to drought stress. Sixteen QTLs were identified in durum wheat by Maccaferri et al. (2008), including two major QTLs on chromosome arms 2BL and 3BS that affected grain yield and showed significant effects in multiple environments (rainfed and irrigated).

Five QTLs for anther-silking interval were identified in the maize drought tolerant line

Ac7643 and transferred to the susceptible line CML247 by marker-assisted backcross. Hybrid lines were obtained that performed better than controls in well watered and mild drought condition in terms of grain yield (Ribaut & Ragot, 2007). The identification of markers or genes associated with root growth and architecture would be particularly useful for breeding programmes to improve root traits by molecular marker-assisted selection. Few papers have described work on the identification of QTLs for root traits in wheat.

Root system architecture (RSA), the spatial configuration of a root system in the soil, is used to describe the shape and structure of root systems. Its importance in plant productivity lies in the fact that major soil resources are heterogeneously distributed in the soil, so that the spatial development of roots will determine the ability of plant to secure edaphic resources (Lynch, 1995). The search for QTLs has been a major research avenue in investigating the genetic variation of RSA, a task that is complicated by the strong responses of RSA to environmental conditions. In several instances overlap of QTLs for root features with those for productivity (yield, water use o capture) has suggested the possible role of the former in determining the latter (Tuberosa et al., 2002a; 2002b; Steele et al., 2007). Although there are few examples of QTLs that individually explained up to 30% of phenotypic variation for root traits in rice (Price & Tomos, 1997) and in maize (Giuliani et al., 2005) and for the response of RSA to environmental factors, root morphology is in most cases regulated by a suite of small-effect loci that interact with the environment (de Dorlodot et al., 2007). This is one of the constrains that limit progress from QTL discovery to the release of new varieties. Some recent papers have reviewed in details the QTLs identified for traits related to drought stress tolerance (Maccaferri et al., 2009; Ashraf, 2010), furthermore, for many crop plants information on drought-related QTL findings have been collected in open source databases, such as GRAMENE (http://www.gramene.org/) or GRAINGENES (http://wheat.pw. usda. gov/GG2/quickquery.shtml#qtls). In particular, Courtois et al. (2009) extracted information from about sixty papers published between 1995 and 2007 and compiled a database containing QTLs for drought tolerance traits and for 29 root parameters. The data describe 2137 root and drought QTLs, out of which 675 for root traits detected in 12 mapping populations.

In rice, several QTLs for root deepness were transferred from the japonica upland cultivar "Azucena", adapted to rainfed conditions, to the lowland indica variety "IR64". MAS selected lines showed a greater root mass in low rainfall trials and higher grain yield (Steele et al., 2007). Following these studies, a highly drought tolerant variety, Birsa Vikas Dhan 111 was released in India, characterized by early maturity, high drought tolerance and high grain yield with good grain quality (Steele, 2009). Linked molecular markers were identified for resistance to cereal cyst nematode (CCN) root disease and the root tolerance to the toxic element Al and are currently used by commercial breeding companies.

Combining, or pyramiding, a number of root characteristics for a target environment can be achieved by phenotype selection in the short term. In future, molecular markers may be available for these characteristics since a gene regulating hydrotropism has been identified in Arabidopsis (Kobayashi et al., 2007); a gene regulating specific root types, including seminal versus nodal roots, has been identified in maize (Taramino et al., 2007); and a significant QTL associated with large root system size was identified in Arabidopsis growing in agar under high osmotic stress (Fitzgerald et al., 2006). Over the past few years there have been several mapping studies that have targeted drought tolerance and other abiotic stress tolerance loci associated with performance in low yielding environments. However, despite this substantial research effort the only markers that have found their way into practical plant breeding programmes are those for boron and aluminium tolerance (Gupta et al., 2010).

MOLECULAR BASES OF PLANT RESPONSE TO WATER STRESS

Molecular and biochemical response of plant to water stress is a very complex task depending on multiple factors (Rizhsky et al., 2002; Bartels & Sourer, 2004). Changes in membrane integrity and modulation of lipid synthesis are key factors in the primary sensing of abiotic stress (Kader & Lindberg, 2010). Secondary, osmotic stress-induced signalling involves changes in plasma membrane H^+-ATPase and Ca^{2+}-ATPase activities that trigger concerted changes of Ca^{2+} influx, cytoplasmic pH, and apoplastic production of ROS (Beffagna et al., 2005).

Transcription factors represent the first level of regulation of mRNA metabolism, controlling the synthesis of pre-mRNA. These molecules are then subject to a splicing process that produces mature mRNA. A well studied phenomenon, with a clear role in regulation of gene expression in stress conditions, is alternative splicing, in which different mRNAs can be produced starting from the same pre-mRNA molecule (Mastrangelo et al., 2011b). The amount of mRNAs in the cell can also be controlled by mechanisms affecting their stability.

Not only proteins but also small non-coding RNA molecules are involved in the regulation of these processes, and they have been recognized as important regulators of gene expression and genome integrity (Ambrosone et al., 2011). Epigenetic regulation, which comprises histone variants and post-translational modifications, DNA methylation and certain small-interfering RNA (siRNA) pathways, controls chromatin structure which can be modified in response to stress. Finally, availability of mRNAs for translation affects the synthesis of the corresponding proteins. In the last years, a new mechanism of posttranscriptional regulation of gene expression was identified in the sequestration of mRNAs in the cytoplasm to generate Stress Granules (SG). SG, produced as result of stress condition, were represented by a subset of mRNAs aggregated with specific proteins, allowing physical separation of these mRNAs from the translational machinery and resulting in transient translational repression (Anderson & Kedersha, 2009).

Transcriptional Factors Influencing the Expression of Genes in Response to Environmental Signals

Plant transcription factors are involved in the response to environmental stresses as critical regulators of the expression of stress-related genes. More than 1,500 genes coding for transcription factors have been annotated in Arabidopsis, and they are classified into several families based on the structure of their DNA-binding domains (Ratcliffe & Riechmann, 2002 – http://datf.cbi.pku.edu.cn/). In particular, members of the MYB, MYC, ERF, bZIP, and WRKY transcription factor families have been implicated in the regulation of plant stress responses (Hussain et al., 2011).

Studies carried out in the model species Arabidopsis allowed to identify different stress signal transduction pathways leading to the activation of members of the above mentioned transcription factor families. These pathways can be either dependent or not by the plant hormone ABA (Hirayama and Shinozaki, 2010).

Among transcription factors depending on ABA, bZIPs are a large family of transcription factors with 75 members annotated in the Arabidopsis genome. Regarding water stress response, the ABRE-binding factor (ABF)/ABA-responsive-element-binding (AREB) proteins respond at the transcriptional and post-transcriptional level to drought and salt stress (Choi et al., 2000; Uno et al., 2000), increasing drought stress toleracence (Table 1). MYC and MYB proteins have a role in late stages of stress response and are also activated following accumulation of endogenus ABA. They generally promote water stress tolerance by acting as positive regulators (Table 1), even if a different mechanism was described for the AtMYB60 and AtMYB44 genes, that are involved in stomatal movements, and function as transcriptional repressors (Cominelli et al., 2005; Jung et al., 2008).

More than 100 members of the NAC gene family have been identified in both Arabidopsis and rice (Fang et al., 2008; Ooka et al., 2003). Members of this family are involved in drought and salinity stress response, as well as in diverse processes as developmental programs, and biotic stress responses (Olsen et al., 2005). RD26, a dehydration-induced NAC protein induced by drought, high salinity, ABA, and JA treatments, represents a key factor in mediating cross-talk between ABA signalling and JA signalling during drought and wounding stress responses (Fujita et al., 2004).

An example of ABA-independent transcription factors acting in drought response are zinc finger homeodomain (ZFHD) proteins. Arabidopsis ZFHD1 binds the ZFHDR motif in the promoter of ERD1 gene, which is also regulated by NAC proteins (Hirayama & Shinozaki, 2010). Ethylene responsive factors (ERFs) represent a class of genes which function in both ABAdependent and independent pathways. They are a transcription factor superfamily that is unique to plants, with 124 members in Arabidopsis (Riechmann et al., 2000). ERF proteins share a conserved 58–59 amino-acid domain (the ERF domain) that binds to two similar ciselements: the GCC box, which is found in several PR (Pathogenesis-Related) gene promoters where it confers ethylene responsiveness, and the C-repeat (CRT)/dehydration-responsive element (DRE) motif, which is involved in the expression of dehydration- and low temperature-responsive genes.

Table 1: Examples of transcription factors regulating drought tolerance in plants

Gene	Gene family	Species	Gene expression	Phenotype of transgenic or mutant plants	Reference
SodERF3		Sugarcane	overexpressed	Improved ABA, Salt and Woundig tolerance (Tobacco)	Trujillo et al., 2008
WXP1		Medicago	overexpressed	Improved Drought tolerance (Arabidopsis)	Zhang et al., 2007
GmERF3	ERF	Soybean	overexpressed	Improved Drought, salt and desease tolerance (Tobacco)	Zhang et al., 2009
RAP2.6		Arabidopsis	overexpressed	Hypersensitive to ABA, salt, osmotic and cold stress (Arabidopsis)	Zhu et al., 2010

DREB1C		Arabidopsis	overexpressed	Enhanced dessication tolerance (Arabidopsis)	Novillo et al., 2004
AtDREB1A		wheat	overexpressed	Delayed wilting under drought stress (Wheat)	Pellegrineschi et al., 2004
AtDREB1A		Tobacco	overexpressed	Improved Drought and cold tolerance (Tobacco)	Kasuga et al., 2004
AtDREB1A		Rice	overexpressed	Improved Drought and salt tolerance (Rice)	Oh et al., 2005
AtCBF4	DRE binding protein 1	Arabidopsis	overexpressed	Improved Drought and freezing tolerance (Arabidopsis)	Haake et al., 2002
OsDREB1		Rice	overexpressed	Improved Drought, Salt and freezing tolerance (Rice)	Ito et al., 2006
HvCBF4		Barley	overexpressed	Increased Drought, Salt and freezing tolerance (Rice)	Oh et al., 2007
AREB1		Arabidopsis	overexpressed	Improved Dehydration survival (Arabidopsis)	Fujita et al., 2005
ABF3/ABF4		Arabidopsis		Improved Drought tolerance (Arabidopsis)	Kang et al., 2002
AREB1		Arabidopsis	overexpressed	Reduced Drought tolerance (Arabidopsis)	Yoshida et al., 2010
ABP9		Maize	overexpressed	Improved photosynthetic capacity under drought stress (Arabidopsis)	Zhang et al., 2008
SlAREB	bZIP	Tomato		Improved Drought and Salt tolerance (Arabidopsis and Tomato)	Hsieh et al., 2010

OsABF1-1, OsABF1-2		Rice		More sensitive to drought and salinity treatments (Rice)	Amir Hossain et al., 2010
OsbZIP23		Rice	overexpressed	Improved Drought and Salt stress tolerance (Rice)	Xiang et al., 2008
WRKY25, WRKY 33		Arabidopsis	overexpressed	Increased sensitivity to ABA and improved salt tolerance (Arabidopsis)	Jiang & Deyholos, 2009
WRKY63	WRKY	Arabidopsis		Decreased drought tolerance and hypersensitive to ABA (Arabidopsis)	Ren et al., 2010
OsWRKY45		Rice	overexpressed	Improved drought tolerance and enhanced desease resistance (Arabidopsis)	Qiu et al., 2009
AtMYB60	TF involved in stomatal movements	Arabidopsis		Decreased wilting under water stress conditions (Arabidopsis)	Cominelli et al., 2005
AtMYB44		Arabidopsis	overexpressed	Improved drought and salt tolerance (Arabidopsis)	Jung et al., 2008
AtMYB15		Arabidopsis	overexpressed	Improved drought tolerance and enhanced sensitivity to ABA (Arabidopsis)	Ding et al., 2009
AtMYB41	R2R3 MYB	Arabidopsis	overexpressed	Negative regulation of transcriptional responses to osmotic stress (Arabidopsis)	Lippold et al., 2009
AtRD26		Arabidopsis	overexpressed	Enhanced sensitivity to ABA (Arabidopsis)	Fujita et al., 2004

ANAC019, ANAC055, ANAC072	NAC	Arabidopsis	overexpressed	Improved drought and salt tolerance (Arabidopsis)	Tran et al., 2004
ONAC045		Rice	overexpressed	Improved drought and salt tolerance (Rice)	Zheng et al., 2009
OsNAC10		Rice	overexpressed	Improved drought tolerance and grain yield (Rice)	Jeong et al., 2010

Therefore, these proteins can have a role in both biotic and abiotic stress responses, as demonstrated for soybean *GmERF3* and the the Arabidopsis ABA-responsive *RAP2.6* genes (Zhang et al., 2009; Zhu et al., 2010). In Arabidopsis, two distinct gene families of DRE/CR binding proteins (*CBF/DREB1* and *DREB2*) were described as two distinct targets of cold and drought ABA-independent signalling transduction pathways, respectively (Shinozahi & Yamaguchi-Shinozaki, 2000). Nevertheless *CBF4*, a member of *CBF/DREB1* family, was described as an ABA-dependent regulator of drought adaptation in Arabidopsis (Haake et al, 2002). *CBF/DREB1* and *DREB2* represent therefore a point of integration of different signal transduction pathways in response to abiotic stresses. The importance of *CBF/DREB* genes for tolerance to abiotic stresses has been well established in particular in cereals, with evidences at level of phenotypic evaluation of over-expressing plants, and co-segregation of *CBF* genes with *QTLs* for frost tolerance (Vàgùjfalvi et al., 2005).

Finally, *WRKY* proteins contain either one or two WRKY domains, 60-amino-acid regions that contain the sequence *WRKYGQK*, and a zinc-finger-like-motif. They are involved in the regulation of diverse plant processes including development, response to various biotic and abiotic stresses, and hormone-mediated pathways (Ramamoorthy et al., 2008). A. thaliana *WRKY25* and *WRKY33* genes are responsive to osmotic stress but they also are regulated by oxidative stress (Miller et al., 2008). Down-stream regulated target genes of *WRKY33* include transcripts with function in ROS detoxification as peroxidases and glutathione-Stransferases (Jiang & Deyholos, 2009).

Stress Related Transcripts from Alternative Splicing Events

Alternative splicing is a process which generates two or more different transcripts from the same pre-mRNA molecule by using different splice sites. The rate of plant genes subject to alternative splicing is comprised between 20 and 70%, depending on the species considered (Mastrangelo et al., 2011b). Alternative splicing events do not randomly affect mRNA of all genes, rather they seem to occur preferentially to mRNAs of certain classes of genes commonly involved in signal transduction, or encoding enzymes, receptors and transcription factors (Ner-Gaon & Fluhr, 2006; Lareau et al., 2004). Four main types of alternative splicing have been described: exon skipping, alternative 5′ and 3′ splice sites and intron retention. The last one is the most common alternative splicing type in plants and fungi (>50% McGuire et al., 2008).

Alternative splicing has been proposed as one of the regulatory mechanisms amplifying the number of proteins that can be produced from a single coding unit. Nevertheless, alternative transcripts containing in frame stop codons, often resulting from retained introns, can be targeted to degradation by nonsense-mediated decay. This mechanism contributes to the fine regulation of the amount of functional protein that will be produced in stressed conditions. Otherwise, truncated polypetides can be produced, which are not necessarily functionless forms of the full length protein. An example is a stress-related transcript of the *MPK13* gene, encoding a protein kinase. This transcript is translated into a truncated protein that has no protein kinase activity, but enhances the *MKK6*-dependent activation of the *MPK13* full-length protein (Lin et al., 2010).

Many of the above described stress-related transcription factors are regulated by alternative splicing. In Arabidopsis, the *AtMYB59* and *AtMYB48* genes were found to code for alternative proteins differing for their MYB repeats and probably for their binding affinities to gene promoters (Li et al., 2006; Fig. 2). A stress-dependent alternative splicing mechanism was described for the *OsDREB2B* gene and its homologs in different species (Mastrangelo et al., 2011b). A transcript containing a shorter ORF (*OsDREB2B1*) accumulated in non stress conditions, and was rapidly converted in the full length transcript

(*OsDREB2B2*) by removal of an exon carrying an in frame stop codon in response to stress exposure. This mechanism is probably aimed to finely and rapidly regulate the amount of functional protein. Moreover, this mechanism can keep the transcription of *OsDREB2B* constitutively active without affecting plant growth (Matsukura et al., 2010).

The serine/arginine proteins are a class of RNA binding proteins involved in splicing regulation. Twenty genes encoding serine/arginine proteins have been identified in Arabidopsis up to now, and most of their mRNAs undergo alternative splicing following developmental and environmental stimuli producing nearly 100 different transcripts (Palusa et al., 2007). They can promote alternative splicing of their own transcripts as well as of other gene products in response to a number of abiotic stresses (Wang & Brendel, 2006). A similar behavior has been shown for some glicyne-rich RNA-binding proteins as *AtGRP7* and *AtGRP8*, which are able to auto-regulate their own splicing and cross-regulate with each other in a negative feed-back loop (Schoning et al., 2008). Alternative splicing regulation of genes producing transcripts that alter the splicing of other genes in turn might considerably enhance and amplify the signal-transduction cascade in response to stress stimuli.

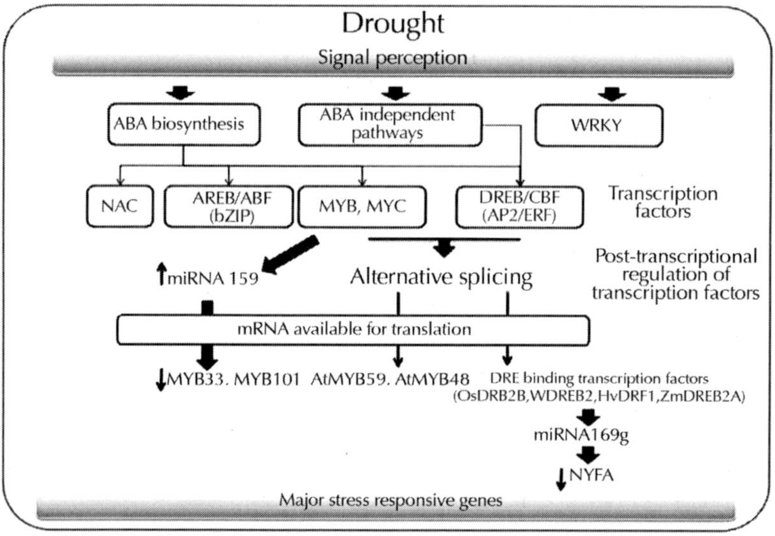

Figure 2: ABA-dependent and independent pathways of response to drought in plants.

Finally, alternative splicing events also have been described for proteins acting in the regulation of gene expression at post-translational level. E3 ubiquitin ligases represent a very large and complex gene family involved in regulation of protein half life by spliceosomemediated protein degradation. Alternative splicing events were described for two Arabidopsis E3 genes (At4g39140 and At2g21500) and for the durum wheat homolog 6G2, whose mRNA retained the last 3′UTR-located intron following exposure to dehydration and cold stress (Mastrangelo et al., 2005). The same stresses induced the accumulation of an alternative transcript for the Arabidopsis SKP1-like 20 (ASK20) gene (Ogura et al., 2008).

Regulatory Proteins Affecting mRNA Availability and Activity

Transport, initiation of translation and degradation by RNA interference have been shown to regulate mRNA levels of genes in response to water stress. Many of these processes are mediated by RNA-binding proteins (RBP), a variety of heterogeneous proteins involved in diverse aspects of post-transcriptional regulation by direct interaction with single/double strand RNA molecules. The processes in which they are involved comprise mRNA maturation events such as splicing, capping, polyadenylation and export from the nucleus. At level of the cytoplasm, they can regulate mRNA localization, stability, decay and translation (Burd & Dreyfuss, 1994; Dreyfuss et al., 2002). RBPs are characterized by conserved RNA-binding motifs, such as RNA recognition, K homology, glycine-rich, arginine-rich, zinc finger (mainly CCCH type - C-x8-C-x5-C-x3-H), and double-stranded RNA-binding motifs. RRM motifs in particular are involved in RNA recognition and in protein–protein interactions, leading to the formation of heterogeneous ribonucleoprotein (RNP) complexes. More than 200 putative RBP genes have been identified in the Arabidopsis and rice genomes, and many of them seem to be unique to plants, suggesting that they might serve plant specific functions (Lorkovic, 2009; Cook et al., 2010).

Expression and/or activity of a number of RBPs were found to be regulated in response to environmental variables, including water deficit, temperature, light and low-oxygen stresses (Park et al., 2009; Sachetto-Martins et al, 2000; Sahi et al., 2007). Several RBPs resulted also to be ABA-regulated, supporting the regulatory role of ABA in the control of posttranscriptional RNA metabolism (Kuhn & Schroeder, 2003).

The role of RBPs in response to drought stress was also demonstrated by using plant mutants. The supersensitive to ABA and drought 1 (sad1) mutant line was isolated in Arabidopsis and exhibited enhanced responses to ABA and drought (Xiong et al., 2001). The ABA hypersensitive 1 (abh1) mutant showed ABA hypersensitive stomatal closing and reduced wilting during drought treatment (Hugouvieux et al., 2001). SAD1 encodes an Smlike protein possibly involved in RNA transport, splicing or degradation, while ABH1 encodes a mRNA cap binding protein which can effectively control ABA signalling components at the RNA level (Covarrubiales & Reyes, 2010). These two genes have been identified as negative regulators of ABA-dependent germination and drought tolerance, together with CBP20 (Cap-Binding Protein 20) and HYL1 (Hyponastic Leaves 1) which codes for a double stranded RNA-binding factor necessary for the biogenesis of miRNAs and crucial for the precise and efficient cleavage of several primary-miRNAs (Vazquez et al., 2004; Szarzynska et al., 2009; Kuhn & Schroeder, 2003).

Some glycine-rich proteins, containing a dispersed CCHC-type zinc finger at the C-terminus (Karlson et al., 2002), have been identified in plants as cold shock domain protein (CSDP) (Verslues et al., 2006). Arabidopsis AtRZ-1a is a cold shock domain protein and has a negative impact on seed germination and seedling growth of Arabidopsis under salt or dehydration stress conditions (Kim et al., 2007). Not only the sequence information, but also the secondary and tertiary structures of RNA molecules contribute to their biological activity. RNA helicases are RBPs that catalyze RNA secondary structure rearrangements, and are potentially required in any cellular process involving RNA maturation (Tanner & Linder, 2001; Rocak & Linder, 2004). The majority of RNA helicases belong to the superfamily 2 (SF2) composed of three subfamilies, termed DEAD, DEAH and DExH/D (Tanner & Linder, 2001). Amino acid sequences outside a common core (Asp-Glu-Ala-Asp) are not conserved and are believed to provide helicase specificity for target RNAs or protein–protein interactions. RNA helicases are associated with a diverse range of biotic cellular functions and are involved in cellular response to abiotic stress. Recently, a temperature-regulated RNA helicase, LOS4, has been linked with developmental processes including flowering and vernalization in Arabidopsis (Gong et al., 2002; 2005). These processes also involve ABA, to which the los4 mutants are sensitive (Gong et al., 2005). Two DEAD-box-related helicases, DNA Helicase 47 (PDH47) and PDH45 are induced by a variety of abiotic stresses in pea (Chinnusamy et al., 2004; Sanan-Mishra et al., 2005; Vashisht et al., 2005). The expression of PDH47 in particular is regulated in a tissue specific manner: the gene is induced by cold and salinity stress in shoots and roots, and by heat and ABA treatment only in roots (Chinnusamy et al., 2004).

Finally, evidences have been reported that helicases can be regulated by the stress-induced alteration of subcellular localization, and by phosphorylation, which provides the opportunity to directly link helicase activity with environmental sensing-signal transduction phosphorylation cascades (Owttrim, 2006).

Degradation of Stress Related Transcript by siRNAs and miRNAs

MicroRNAs (miRNAs) and siRNAs are small noncoding RNAs that have recently emerged as important regulators of mRNA degradation, translational repression, and chromatin modification. miRNAs form an abundant class of tiny RNAs characterized by a high level of conservation across species, suggesting a common evolutionary basis. They act in regulating the expression of protein-coding genes in multicellular eukaryotes (Bartel, 2004). Plant miRNAs participate in numerous processes, including development, pattern formation, flowering time, hormone regulation, nutrient limitation, response to stress, and even self-regulation of the miRNA biogenesis pathway (Jones-Rhoades et al., 2006). Regarding their involvement in stress response, abiotic stresses like cold, dehydration, salt stress and nutrient starvation regulate the expression of different plant miRNAs (Lu & Huang, 2008). An example is the Arabidopsis miR393, that is up-regulated by cold, dehydration, high salinity, and abscisic acid (ABA) treatments (Sunkar & Zhu, 2004). In maize 21 miRNA differentally expressed under drought stress were identified (Chen et al., 2010).

In order to understand the mechanisms by which they exert a role in stress protection, it is important to characterize their target mRNAs. At this regard, an interesting feature of miRNAs is the fact that their targets are often regulatory genes (Jones-Rhoades & Bartel, 2004; Rhoades et al., 2002; Zhang et al., 2006). The level of miR159 increased in Arabidopsis seedlings water stressed. In arabidopsis transgenic plants the over-expression of miRNA159 reduced the level of MYB33 and MYB101 transcripts, and a hyposensitive phenotype to ABA was observed (Reyes & Chua, 2007; Fig. 2).

Sunkar and Zhu (2004) reported other ABA induced miRNAs (miR397b and miR402) but also cases of miRNA down-regulated by this hormone (miR389a). Two members of the miR169 gene family, miR169a and miR169c, are repressed following drought treatments in Arabidopsis. As their target is the nuclear factor Y transcription factor NFYA5, the aboundance of this transcript increases and promotes stress response in mature plants (Li et al., 2008). Nevertheless, even if the same conserved miRNA family regulates homologous targets in two different plant species, the effects of this regulation can be different. Members of the miR169 family in rice, miR169g and miR169n/o are induced by salt (Zhao et al., 2009) and drought (Zhao et al., 2007) and differences in levels of induction can be observed in different tissues, being more prominent in roots than in shoots. Interestingly, miR-169g, that acts reducing the expression of NFYA, may be regulated directly by DREB transcriptional factors (Zhao et al., 2009).

Ten percent of Arabidopsis genes are in convergent overlapping gene pairs, also known as natural cis-antisense gene pairs and overlapping transcripts in antisense orientation could form double-stranded RNAs that may be processed into small RNAs (Jen et al., 2005; Wang et al., 2005). These nat-siRNAs (natural antisense transcripts-generated siRNAs) have recently emerged as important players in plant stress responses. A study in Arabidopsis demonstrated the involvement of nat-siRNA in the accumulation of proline during response to stress. As an example, the Arabidopsis P5CDH gene, involved in proline catabolism, is down-regulated in response to salt stress following the induction of SRO5, a gene of unknown function. The two genes form an antisense overlapping gene pair that generates two siRNAs (Borsani et al., 2005).

Epigenetic Contribution to Water Stress Response in Plants

Epigenetic regulation is emerging as an important mechanism in response to stress. Drought induced linker histone variant H1-S was shown to be involved in the negative regulation of stomatal conductance based on the phenotypic analysis of antisense transgenic H1-S tomato plants (Scippa et al., 2004). Several hystone deacetylases (HDACs) are induced by ABA in rice (Fu et al., 2007) and Arabidopsis (Sridha et al., 2006). Transgenic Arabidopsis plants overexpressing AtHD2C exhibited enhanced expression of ABA-responsive genes and greater salt and drought tolerance than the WT plants (Sridha et al., 2006). Besides acetylation and de-acetylation, other post-translational mechanisms can regulate the abundance and activity of histones. In particular, histone phosphorylation and ubiquitination showed a role in enhancing gene transcription (Sridhar et al., 2007; Zhang et al., 2007), while biotinylation and sumoylation repress gene expression (Nathan et al., 2006; Camporeale et al., 2007). In the desert shrub *Zygophyllum dumosum* methylation level of histone H3 was higher in presence of water than under dry growth conditions indicating post-translational regulation of gene expression activity (Granot et al., 2009). ABA-mediated pathways also are involved in epigenetic modifications, as suggested by the ABA-dependent regulation of barley Polycomb proteins expression, with a role in histone methylation control (Kapazoglou et al., 2010).

Studies on Arabidopsis over-expressing or knock out lines for the SNF2/ BRAHMA-type chromatin remodeling gene AtCHR12 indicated a role of this gene in regulation of growth, in particular under drought and heat stresses (Mlynarova et al., 2007). In *Pisum sativum* ABA and drought stress induced the expression of the chromatin remodelling PsSNF5 gene. PsSNF5 protein interacts with Arabidopsis SWI3-like proteins (SWI3A and SWI3B), which in turn interact with FCA, a protein involved in the regulation of flowering (Sarnowski et al., 2005; Rios et al., 2007). This is a clear example in which stress response and plant development are co-ordinately regulated through chromatin remodeling. Direct DNA methylation can also be involved in plant stress response. Drought and salt stresses induced a switch in photosynthesis mode from C3 to CAM in the facultative halophyte

Mesembryanthemum crystallinum L. This metabolic change was associated with stress-induced-specific CpHpG-hypermethylation of satellite DNA (Dyachenko et al., 2006).

In natural populations of mangroves DNA was hypomethylated when grown under saline conditions in contrast to populations from non-saline sites (Lira- Medeiros et al., 2010). Although global analysis in plants such as Arabidopsis and rice suggests that the vast majority of transposons are inactive, methylated, and targeted by siRNAs (Nobuta et al., 2007; Lister et al., 2008), the induction of alternative epigenetic states not only triggers the formation of novel epialleles but also promotes the movement of DNA transposons and retroelements that are very abundant in plant genomes (Reinders et al., 2009; Mirouze et al., 2009). A lot of examples of environmentally induced transposon activities were reported (Slotkin & Martienssen, 2007), as the family of copia retrotrasposon, named Onsen, activated by heat stress in Arabidopsis (Ito et al., 2011). In natural populations, stress may play a role in transposon amplification. An example is the copy number of BARE-1 retrotransposons in barley, which varies in natural populations depending on aridity of growth environment (Vicient et al., 1999). These evidences indicate that plant populations living in stressed environments may carry inherited memories of stress adaptation and transfer this epigenetically to next generations.

THE MOLECULAR RESPONSE OF PLANTS TO WATER STRESS: A COMPLEX FRAME RESULTING FROM INTEGRATION OF MULTIPLE REGULATION LAYERS

The plant response mechanisms to water deficit strictly depend on plant developmental stage, stress intensity and stress duration (Bartels & Souer 2004; De Leonardis et al., 2007). A study on 325 rice transcription factors demonstrated that many of them have a tissue or developmental stage specific expression (Duan et al., 2005). In a genome wide study, Bray (2004) compared three independent array experiments dedicated to the Arabidopsis water stress response. The experiments differed for plants age, substrate of growth and stress applications. Only a small set of genes were commonly induced or repressed. Similar results were obtained in wheat, barley and rice (Ozturk et al., 2002; Lan et al., 2005; Mohammadi et al., 2007). In Arabidopsis the *Nine-Cisepoxycarotenoid Dioxygenase* 3 (NCED3), DREB2A and RD29B genes were expressed with different levels and timing following two different kinds of stress imposition (Harb et al., 2011). These differences observed following a rapid or gradual water stress are probably due to the need of plants to optimally react to a stress event as it occurs in field conditions. Therefore, many internal and external stimuli have to be integrated into common signalling pathways. Moreover, plants usually are exposed in field to different kinds of stress simultaneously, and the effect of the combined stresses in terms of gene expression is not simply the sum of the effects produced by the stresses applied separately (Rizhsky et al., 2002). The superimposed complexity levels in the response to environmental changes, are therefore aimed to ensure temporally and spatially appropriate patterns of downstream stress-related gene expression.

After the translation, many post-translational mechanisms can target proteins modifying their activity, sub-cellular localization and half-life (Downes & Vierstra, 2005). Phosphorylation is one of the best known mechanism that plays a key role in many biological processes, as phosphorylation/de-phosphorylation cascades commonly translate extracellular stimuli into the activation of specific responses (Boudsocq & Laurière, 2005). Among the polypeptides, ubiquitin and SUMO conjugations are emerging as major posttranslational regulatory processes in all eukaryotes (Stone & Callis, 2007; Miura et al., 2007).

The covalent binding of poly-ubiquitin usually targets proteins for proteolysis. Conversely, monoubiquitination regulate the location and activity of proteins, affecting various cellular processes from transcriptional regulation to membrane transport (Hicke, 2001). Similar effects are produced by the covalent conjugation of the Small Ubiquitin-like Modifier (SUMO) peptide (Hay et al., 2005). Both ubiquitination and sumoylation are involved in the promotion of stress tolerance in plants, and they offer a very clear example of multiple layer control of key regulators of the stress response. Along this chapter DREB/CBF proteins have been described as transcription factors with a pivotal role in plant tolerance to cold and drought stress. Their expression has shown to be modulated at transcriptional level, but also by alternative splicing (Matsukura et al., 2010). Furthermore, the HOS1 protein, correspondending to an E3 ubiquitin ligase, mediates the ubiquitination of the master regulator for the response to cold, the transcription factor Inducer of CBF Expression 1, ICE1, and repressor of MYB15 expression. This leads to its proteasome-mediated degradation during exposure to cold (Dong et al., 2006). ICE1 protein, in turn, is stabilized by sumoylation that therefore acts in this pathway with an antagonistic role with respect to ubiquitination (Ishitani et al., 1998). Finally, DREB transcriptional factors can down-regulate the expression of NFYA through activation of miRNAs (Zhao et al., 2009).

The great complexity of the pathway of regulation of gene expression in plant response to water stress makes the analysis of transcriptome in different conditions not suitable alone to draw a clear picture of tolerance mechanisms. Variations at level of proteins and ultimately of metabolites have to be investigated to achieve a more complete evaluation. In this light, recent advances in profiling of plant proteome and metabolome in water stress conditions have provided chances to integrate data from gene expression and protein activities studies. Outcomes indicate an important role of post-transcriptional and post-translational mechanisms in coordinating the plant molecular response to water stress (Mazzucotelli et al., 2008).

THE CONTRIBUTION OF GENETIC AND MOLECULAR KNOWLEDGE TO THE IMPROVEMENT OF DROUGHT TOLERANCE IN FIELD

A very complex network of gene interactions in response to water stress has been described in the last years, and a high number of QTLs, widespread in the genome, have been identified for tolerance, each of them controlling a low percentage of explained phenotypic variabilty. In some cases the molecular basis of resistance QTLs has been explained. Genes having a role in stress tolerance were shown to co-localize with tolerance QTLs in mapping populations. This is the case of DREB/CBF genes, for which a large gene cluster has been mapped in correspondence of QTLs for frost and drought tolerance in barley and wheat (Vàgùjfalvi et al., 2005; Francia et al., 2007). On the other side, many efforts are in course to isolate the gene(s) behind tolerance QTLs, in order to have access to the transgenic approach, or to design perfect molecular markers to pyramid different QTLs into the same genotype through MAS without the risk of losing association due to recombination. The Arabidopsis ERECTA gene was cloned, as the sequence beyond a QTL for transpiration efficiency (Masle et al., 2005).

Even if the molecular basis of QTLs is not known, some examples are available in which the transfer of some tolerance QTLs in MAS programs has contributed to increase grain yield in water stress conditions (see paragraph n. 3).

Anyway, the investigation of molecular mechanisms which concur in regulating the water stress response in plant allows the identification of genes/processes with a key role in determining tolerance. The expression of these genes can be altered in transgenic plants in order to obtain a tolerant phenotype. Besides the genes reported in Table 1, in some cases this approach has been successful in increasing agronomic performance of plants in the field. An example is represented by transgenic wheat constitutively expressing the barley HVA1 gene, encoding a member of the group 3 late embryogenesis abundant (LEA) proteins. Results of nine field experiments over six cropping seasons, showed that the HVA1 protein confers a significant protection from water stress (Bahieldin et al., 2005). Aquaporins mediate most of the symplastic water transport in plants, which represents a limiting factor for plant growth and vigor in particular under unfavorable growth conditions and abiotic stress. Differential expression of genes that encode different aquaporin isoforms during plant development has been shown to be associated with various physiological processes. Such processes include stomatal closure and opening, organ movement, cell elongation, and cell division (Kaldenhoff et al., 2008). The SlTIP2 gene coding for an aquaporin was particularly effective in improving water stress resistance of tomato plants (Sade et al., 2009). Another successful gene is OsNAC10, introduced in field-tested rice plants under the control of the constitutive promoter GOS2 and the root-specific promoter RCc3 (Jeong et al., 2010).

PERSPECTIVES

Two different but complementary approaches have been presented in this chapter for the improvement of water stress tolerance. In the first one, the phenotypic and molecular evaluation of suitable genetic materials leads to the identification of genomic regions involved in the control of tolerance. At the same time, closely linked molecular markers are found, which can be used in MAS programs to transfer useful alleles for tolerance. In the second one, the molecular study of the water stress response in plant leads to the identification of genes/processes with a key role in determining tolerance.

In the last years, strong technical advances have been realized, in the frame of the "omic" technologies, which make the study of genomes, transcriptomes, proteomes, metabolomes and phenomes more rapid and precise. Methods for a more fine phenotypic evaluation of a high number of individuals, in both controlled and field conditions, are needed for an accurate genetic analysis on segregating populations or germplasm collections. The development in particular of new DNA sequencing technologies rapidly is producing huge amounts of sequence information with a number of applications including genome resequencing and polymorphism detection, mutation mapping, DNA methylation and histone modification studies, transcriptome sequencing, gene discovery, alternative splicing identification, small RNA profiling and DNA-protein interactions (Lister et al., 2008; Delseny et al., 2010). Thanks to these advancements, new perspectives are open for the investigation of genetic and molecular basis of water stress tolerance. Sequencing of entire genomes of crop species is expected to provide a huge opportunity to clone QTLs for drought-related traits in the near future. Moreover, sequence analysis on a genome-wide scale allows the fast and low-cost development of extremely high number of molecular markers. The availability in particular of large Single Nucleotide Polymorphism (SNP) panels for crops will accelerate the QTL discovery and transfer in MAS programs already in course for single marker-trait associations. Furthermore, it will be possible to apply a new method called genomic selection (Meuwissen et al., 2001), which predicts breeding values using data deriving from all molecular markers covering the whole genome at the same time. In this way, breeders now have an opportunity to integrate classical phenotype-based selection with selection on the basis of genotype. In particular, they will have the possibility to follow genomic variations associated to many traits of interest at the same time.

ACKNOWLEDGEMENTS

This study was supported by the Italian Ministry of Agriculture (MiPAAF), with the special grants AGRONANOTECH and MAPPA 5A, and by the Ministry of Education, University and Research (MIUR), with the special grant AGROGEN.

REFERENCES

1. Abeledo, L.G., Calderoni, D.F. & Slafer, G.A. (2002). Genetic improvement of barley yield potential and its physiological determinants in Argentina (1944–1998), Euphytica Vol.130: 325-334.
2. Ambrosone, A., Costa, A., Leone, A. & Grillo, S. (2011). Beyond transcription: RNA-binding proteins as emerging reguators of plant response to environmental constrains, Plant Science doi: 10.1016/J.plantsci. 2011.02.004.
3. Amir Hossain, M., Lee, Y., Cho, J.I., Ahn ,C.H., Lee, S.K., Jeon, J.S., Kang, H., Lee, C.H., An, G. & Park, P.B. (2010), The bZIP transcription factor OsABF1 is an ABA responsive element binding factor that enhances abiotic stress signaling in rice, Plant Molecular Biology Vol.72: 557–566.
4. Amudha, J., & Balasubramani, G. (2011). Recent molecular advances to cambat abiotic stress tolerance in crop plants, Biotechnology and Molecular Biology Review Vol.6(No. 2): 31–58.
5. Anderson, P. & Kedersha, N. (2009). RNA granules: post-transcriptional and epigenetic modulators of gene expression, Naural. Review Molecular Cell Biology Vol.10: 430-436.
6. Araus, J.L., Slafer, G.A., Reynolds, M.P. & Royo, C. (2002). Plant breeding and drought in C-3 cereals: what should we breed for?, Annals of Botany Vol.89: 925–940.

7. Ashraf, M. (2010). Inducing drought tolerance in plants: recent advances, Biotechnol Advances Vol. 28(No.1): 169–183.
8. Bahieldin, A., Mahfouz, H.T., Eissa, H.F., Saleh, O.M., Ramadan, A.M., Ahmed, I.A., Dyer, W.E., El-Itriby, H.A. & Madkour, M.A. (2005). Field evaluation of transgenic wheat plants stably expressing the HVA1 gene for drought tolerance, Physiologia Plantarum Vol. 123: 421-427.
9. Bartels, D. & Sourer, E. (2004). Molecular responses of higher plants to dehydration, in H. Hirt, K. Shinozaki (ed.), Plant Responses to Abiotic Stress, Springer Verlag, Berlin, Heildelberg, pp. 9–38.
10. Bartel, D.P. (2004). MicroRNAs: Genomics, biogenesis, mechanism, and function, Cell Vol.116: 281–297.
11. Beffagna, N., Buffoli, B. & Busi, C. (2005). Modulation of reactive oxygen species production during osmotic stress in Arabidopsis thaliana cultured cells: involvement of the plasma membrane Ca2+- ATPase and H+ ATPase, Plant Cell Physiology Vol. 46 :1326–1339.
12. Bengough, A.G., McKenzie, B.M., Hallett, P.D. & Valentine, T.A. (2011). Root elongation, water stress, and mechanical impedance: a review of limiting stresses and beneficial root tip traits, Journal of Experimental Botany Vol.62 (No. 1). 59–68.
13. Bolanos, J. & Edmeades, G.O. (1996). The importance of anthesis-silking interval in breeding for drought tolerance in tropical maize, Field Crops Research Vol.48(No. 1): 65-80.
14. Borsani, O., Zhu, J., Verslues, P.E., Sunkar, R. & Zhu, J.-K. (2005). Endogenous siRNAs derived from a pair of natural cis-antisense transcripts regulate salt tolerance in Arabidopsis, Cell Vol.123: 1279–1291.
15. Boudsocq, M. & Laurière, C. (2005). Osmotic signaling in plants: multiple pathways mediated by emerging kinase families, Plant Physiology Vol.138: 1185–1194.
16. Bray, E.A. (2004). Genes commonly regulated by water-deficit stress in Arabidopsis thaliana, Journal Experimental Botany Vol.55: 2331–2341.
17. Burd, C.G. & Dreyfuss, G. (1994). Conserved structures and diversity of functions of RNAbinding proteins, Science Vol.265: 615–621.

18. Calderini, D.F. & Slafer, G.A. (1998). Changes in yield and yield stability in wheat during the 20th century, Field Crops Reseach Vol.57: 335-347.

19. Camporeale, G., Oommen, A.M., Griffin, J.B., Sarath, G. & Zempleni, J. (2007). K12-biotinylated histone H4 marks heterochromatin in human lymphoblastoma cells, The Journal of Nutritional Biochemistry Vol. 18: 760-768.

20. Cattivelli, L., Delogu, G., Terzi, V. & Stanca, A.M. (1994). Progress in barley breeding, in Slafer GA (ed.), Genetic Improvement of Field Crops, Marcel Dekker, Inc. New York, pp. 95-181.

21. Cattivelli, L., Rizza, F., Badeck, F.-W., Mazzucotelli, E., Mastrangelo, A.M., Francia, E., Marè, C., Tondelli, A. & Stanca, A.M. (2008). Drought tolerance improvement in crop plants: an integrated view from breeding to genomics, Field Crops Reseach Vol.105: 1-14.

22. Chaves, M.M., Pereira, J.S., Maroco, J., Rodrigues, M.L., Ricardo, C.P.P., Osòrio, M.L., Carvalho, I., Faria, T. & Pinheiro, C. (2002). How Plants Cope with Water Stress in the Field? Photosynthesis and Growth, Annals of Botany Vol.89(No. 7): 907–916.

23. Chen, X., Yang, R.-F., Li, W.-C. & Fu, F.-L. (2010). Identification of 21 microRNAs in maize and their differential expression under drought stress, African Journal of Biotechnology Vol. 9 (No. 30): 4741-4753.

24. Chinnusamy, V., Schumaker, K. & Zhu, J.K. (2004). Molecular genetic perspectives on crosstalk and specificity in abiotic stress signalling in plants, Jornal Experimental Botany Vol.55: 225–236.

25. Choi, H., Hong, J., Ha, J., Kang, J. & Kim, S.Y.(2000). ABFs, a family of ABA-responsive element binding factors, Journal of Biological Chemistry Vol. 275:1723-1730.

26. Ciais, P., Reichstein, M., Viovy, N., Granier, A., Ogèe, J. et al. (2005). Europe-wide reduction in primary productivity caused by the heat and drought in 2003, Nature Vol.472: 529-533.

27. Cominelli E., Galbiati M., Vavasseur A., Conti L., Sala T., Vuylsteke M., Leonhardt N., Dellaporta S. & Tonelli C. (2005). A guard-cell-specific MYB transcription factor regulates stomatal movements and plant drought tolerance, Current Biology Vol.15:1196–1200.

28. Cook, K.B., Kazan, H., Zuberi, K., Morris, Q. & Hughes, T.R. (2010). RBPDB: a database of RNA-binding specificities, Nucleic Acids Research Vol.39(suppl.1): 301-308.
29. Courtois, B., Ahmadi, N., Khowaja, F., Price, A.H., Rami, J.F., Frouin, J., Hamelin, C. & Ruiz, M. (2009). Rice Root Genetic Architecture: Meta-analysis from a Drought QTL Database, Rice Vol.2: 115–128.
30. Covarrubiales, A.A. & Reyes, J.L. (2010). Post-transcriptional gene regulation of salinity and drought responses by plant microRNAs, Plant Cell and Environment Vol.33: 481–489.
31. de Dorlodot, S., Forster, B., Pagè, L., Price, A., Tuberosa, R. & Draye, X. (2007). Root system architecture: opportunities and constraints for genetic improvement of crops, Trends in Plant Science Vol.12: 474–481.
32. De Leonardis, A. M., Marone, D., Mazzucotelli, E., Neffar, F., Rizza, F., Di Fonzo, N., Cattivelli, L. & Mastrangelo, A.M. (2007). Durum wheat genes up-regulated in the early phases of cold stress are modulated by drought in a developmental and genotype dependent manner, Plant Science Vol.172: 1005–1016.
33. De Vita, P., Mastrangelo, A.M., Matteu, L., Mazzucotelli, E., Virzì, N., Palumbo, M., Lo Storto, M., Rizza, F. & Cattivelli, L. (2010). Genetic improvement effects on yield stability in durum wheat genotypes grown in Italy, Field Crops Research Vol.119: 68–77.
34. Delseny, M., Han, B. & Hsing, Y.I. (2010). High throughput DNA sequencing: The new sequencing revolution, Plant Science Vol. 179: 407-422.
35. Dichio, B., Xiloyannis, C., Sofo, A. & Montanaro, G. (2006). Osmotic regulation in leaves and roots of olive tree (Olea europaea L.) during water deficit and rewatering. Tree Physiology, Vol. 26: 179–185.
36. Ding, Z., Li, S., An, X., Liu, X., Qin, H., & Wang, D. (2009). Transgenic expression of MYB15 confers enhanced sensitivity to abscisic acid and improved drought tolerance in Arabidopsis thaliana, Journal of Genetics & Genomics Vol.36: 17–29.
37. Dong, C.H., Agarwal, M., Zhang, Y., Xie, Q. & Zhu, J.K. (2006). The negative regulator of plant cold responses, HOS1, is a RING E3 ligase that mediates the ubiquitination and degradation of ICE1, Proc. Natl. Acad. Sci. U.S.A. Vol.103: 8281–8286.

38. Downes, B. & Vierstra, R.D. (2005). Post-translational regulation in plants employing a diverse set of polypeptide tags, Biochemical Society Transaction Vol.33: 393–399.

39. Dreyfuss, G., Kim, V.N. & Kataoka, N. (2002). Messenger-RNA-binding proteins and the messages they carry, Nature Reviews Molecular Cell Biology Vol.3: 195–205.

40. Duan, K., Luo, Y.-H., Luo, D., Xu, Z.-H. & Xue, H.-W. (2005). New insights into the complex and coordinated transcriptional regulation networks underlying rice seed development through cDNA chip-based analysis, Plant Molecular Biology Vol.57: 785–804.

41. Dyachenko, O.V., Zakharchenko, N.S., Shevchuk, T.V., Bohnert, H.J., Cushman, J.C. & Buryanov, Y.I. (2006). Effect of hypermethylation of CCWGG sequences in DNA of *Mesembryanthemum crystallinum* plants on their adaptation to salt stress, Biochemistry (Moscow) Vol.71: 461-465.

42. Eapen, D., Barroso, M.L., Ponce, G., Campos, M.E. & Cassab, G.I. (2005). Hydrotropism: Root growth responses to water, Trends in Plant Science Vol.10: 44–50.

43. Fang, Y., You, J., Xie, K., Xie, W. & Xiong, L. (2008). Systematic sequence analysis and identification of tissue-specific or stress-responsive genes of NAC transcription factor family in rice, Molecular Genetics & Genomics Vol.280: 547–563.

44. Farquhar, G.D. & Richards, R.A. (1984). Isotopic composition of plant carbon correlates with water-use efficiency of wheat genotypes, Australian Journal of Plant Physiology Vol.11: 539–552.

45. Fischer, R.A., Rees, D., Sayre, K.D., Lu, Z.-M., Condon, A.G. & Larque Saavedra, A. (1998). Wheat Yield Progress Associated with Higher Stomatal Conductance and Photosynthetic Rate, and Cooler Canopies, Crop Science Vol. 38 (No. 6): 1467-1475.

46. Fitzgerald, J.N., Lehti-Shui, M.D., Ingram, P.A., Deak, K.I., Biesiada, T. & Malamy, J.E. (2006). Identification of quantitative trait loci that regulate Arabidopsis root system size and plasticity, Genetics Vol.172: 485–498.

47. Francia, E., Barabaschi, D., Tondelli, A., Laidò, G. Rizza, F., Stanca, A.M., Busconi, M. & Fogher, C. (2007). Fine mapping of

a HvCBF gene cluster at the frost resistance locus Fr-H2 in barley, Theoretical and Applied Genetics Vol.115: 1083–1091.

48. Fu, W., Wu, K. & Duan, J. (2007). Sequence and expression analysis of histone deacetylases in rice, Biochemical and Biophysical Research Communication Vol.356: 843-850.

49. Fujita. M., Fujita, Y., Maruyama, K., Seki, M., Hiratsu, K., Ohme-Takagi, M., Tran, L.S.P., Yamaguchi-Shinozaki, K. & Shinozaki, K. (2004). A dehydration-induced NAC protein, RD26, is involved in a novel ABA-dependent stress-signaling pathway The Plant Journal Vol.39: 863–876.

50. Fujita, Y., Fujita, M., Satoh, R., Maruyama, K., Parvez, M., Seki, M., Hiratsu, K., OhmeTakagi, M., Shinozaki, K. & Yamaguchi-Shinozaki, K. (2005). AREB1 is a transcriptional activator of novel ABRE dependent ABA signaling that enhances drought stress tolerance in Arabidopsis, Plant Cell Vol.17: 3470–3488.

51. Garrity, D.P. & O'Toole, J.C. (1994). Screening rice for drought resistance at the reproductive phase, Field Crops Research Vol.39: 99-110.

52. Giuliani, S., Sanguineti, M.C., Tuberosa, R., Bellotti, M., Salvi, S. & Landi, P. (2005). RootABA1, a major constitutive QTL, affects maize root architecture and leaf ABA concentration at different water regimes, Journal of Experimental Botany Vol.56: 3061–3070.

53. Gong, Z., Dong, C.H., Lee, H., Zhu, J., Xiong, L., Gong, D., Stevenson, B. & Zhu, J.K. (2005). A DEAD box RNA helicase is essential for mRNA export and important for development and stress responses in Arabidopsis, Plant Cell Vol.17: 256–267.

54. Gong, Z., Lee, H., Xiong, L., Jagendorf, A., Stevenson, B. & Zhu, J.K. (2002). RNA helicaselike protein as an early regulator of transcription factors for plant chilling and freezing tolerance, Proc. Natl Acad. Sci. USA Vol.99: 11507–11512.

55. Granot, G., Sikron-Persi, N., Gaspan,O., Florentin, A., Talwara, S., Paul, L.K., Morgenstern, Y., Granot, Y. & Grafi, G. (2009). Histone modifications associated with drought tolerance in the desert plant *Zygophyllum dumosum* Boiss, Planta Vol.231: 27–34.

56. Gupta, P., Langridge, P. & Mir, R.R. (2010). Marker-assisted

wheat breeding: present status and future possibilities, Molecular Breeding Vol.26: 145–161.

57. Haake, V., Cook, D., Riechmann, J.L., Pineda, O., Thomashow, M.F. & Zhang J.Z. (2002). Transcription factor CBF4 is a regulator of drought adaptation in Arabidopsis, Plant Physiology Vol.130: 639–648.
58. Harb, A., Krishnan, A., Ambavaram, M.M.R. & Pereira, A. (2011). Molecular and Physiological Analysis of Drought Stress in Arabidopsis Reveals Early Responses Leading to Acclimation in Plant Growth, Plant Physiology Vol.154: 1254–1271.
59. Hay, R.T. (2005). SUMO: a history of modification, Molecular Cell Vol.18: 1–12.
60. Hicke, L. (2001). Protein regulation by monoubiquitin, Nature Reviews Molecular Cell Biology Vol.2: 195–201.
61. Hirayama, T. & Kazuo Shinozaki, K.(2010). Research on plant abiotic stress responses in the post-genome era: past, present and future, The Plant Journal Vol.61: 1041–1052.
62. Ho, M.D., Rosas, J.C., Brown, K.M. & Lynch, J.P. (2005). Root architectural tradeoffs for water and phosphorus acquisition, Functional Plant Biology Vol.32: 737–748.
63. Hocholdinger, F., Park, W.J., Sauer, M. & Woll, K. (2004). From weeds to crops: Genetic analysis of root development in cereals, Trends in Plant Science Vol.9: 42–48.
64. Hsieh, T.H., Li, C.W., Su, R.C., Cheng, C.P., Sanjaya, T. et al., (2010). A tomato bZIP transcription factor, SlAREB, is involved in water deficit and salt stress response, Planta Vol.231: 1459–1473.
65. Hugouvieux, V., Kwak, J.M. & Schroeder, J.I. (2001). An mRNA cap binding protein, ABH1, modulates early abscisic acid signal transduction in Arabidopsis, Cell Vol.106: 477–487.
66. Hussain, S.S., Kayani, M.A. & Amjad, M. (2011). Transcription Factors as Tools to Engineer Enhanced Drought Stress Tolerance in Plants, American Institute of Chemical Engineers February Vol.7: 297-306.
67. Ingram, J. & Bartels D. (1996). The molecular basis of dehydration tolerance in plants. Annual Reviews Plant Physiology Plant Molecular Biology Vol.47: 377–403.
68. Ishitani, M., Xiong, L., Lee, H., Stevenson, B. & Zhu, J.K.

(1998). HOS1, a genetic locus involved in cold-responsive gene expression in Arabidopsis, Plant Cell Vol.10: 1151–1162.

69. Ito, Y., Katsura, K., Maruyama, K., Taji, T., Kobayashi, M., Seki, M., Shinozaki, K. & Yamaguchi-Shinozaki, K. (2006). Functional analysis of rice DREB1/CBF-type transcription factors involved in cold-responsive gene expression in transgenic rice, Plant Cell Physiology Vol.47: 141–153.
70. Ito, H., Gaubert, H., Bucher, E., Mirouze, M., Vaillant, I. & Paszkowski, J. (2011). An siRNA pathway prevents transgenerational retrotransposition in plants subjected to stress, Nature Vol.472: 115-119.
71. Jen, C.-H., Michalopoulos, I., Westhead, D.R. & Meyer, P. (2005). Natural antisense transcripts with coding capacity in Arabidopsis may have a regulatory role that is not linked to double-stranded RNA degradation, Genome Biology Vol.6: R51.
72. Jeong, J.S., Kim, Y.S., Baek, K.H., Jung, H., Ha, S.H., Do Choi, Y., Kim, M., Reuzeau, C. & Kim, J.K. (2010). Root-specific expression of Os-NAC10 improves drought tolerance and grain yield in rice under field drought conditions, Plant Physiology Vol.153: 185–197.
73. Jiang, Y. & Deyholos. M.K. (2009). Functional characterization of Arabidopsis NaClinducible WRKY25 and WRKY33 transcription factors in abiotic stresses, Plant Molecular Biology Vol.69: 91–105.
74. Jones-Rhoades, M.W. & Bartel, D.P. (2004). Computational identification of plant microRNAs and their targets, including a stress-induced miRNA, Molecular Cell Vol.14: 787-799.
75. Jones-Rhoades, M.W., Bartel, D.P. & Bartel, B. (2006). MicroRNAs and their regulatory roles in plants, Annual Review of Plant Biology Vol. 57: 19–53.
76. Jung, C., Seo, J.S., Han, S.W., Koo, Y.J., Kim, C.H., Song, S.I., Nahm, B.H., Choi, Y.D. & Cheong, J.J. (2008). Overexpression of AtMYB44 enhances stomatal closure to confer abiotic stress tolerance in transgenic Arabidopsis, Plant Physiology Vol.146: 623–635.
77. Kader, M.A. & Lindberg, S. (2010). Cytosolic calcium and pH signaling in plants under salinity stress, Plant Signaling & Behaviour Vol. 5: 233–238.

78. Kaldenhoff, R., Ribas-Carbo, M., Flexas Sans, J., Lovisolo, C., Heckwolf, M. & Uehlein, N. (2008). Aquaporins and plant water balance, Plant, Cell and Environment Vol.31: 658–666.

79. Kang, J., Choi, H., Im, M. & Kim, S.Y. (2002). Arabidopsis basic leucine zipper proteins that mediate stress-responsive abscisic acid signalling, Plant Cell Vol.14: 343-357.

80. Kapazoglou, A., Tondelli, A., Papaefthimiou, D., Ampatzidou, H., Francia, E., Stanca, M.A., Bladenopoulos, K. & Tsaftaris, A.S. (2010). Epigenetic chromatin modifiers in barley: IV. The study of barley polycomb group (PcG) genes during seed development and in response to external ABA, BMC Plant Biology Vol.10: 73.

81. Karlson, D., Nakaminami, K., Toyomasu, T. & Imai, R. (2002). A cold-regulated nucleic acidbinding protein of winter wheat shares a domain with bacterial cold shock proteins, The Journal of Biological chemistry Vol.277: 35248–35256.

82. Kasuga, M., Miura, S. & Yamaguchi-Shinozaki, K. (2004). A combination of the Arabidopsis DREB1A gene and stress inducible rd29A promoter improved drought and low temperature stress tolerance in tobacco by gene transfer, Plant Cell Physiology Vol.45: 346–350.

83. Kerstiens, G. (2006). Water transport in plant cuticles: an update, Journal of Experimental Botany Vol. 57(No. 11): 2493-2499.

84. Kim, Y.O., Pan, S.O., Jung, C.H. & Kang, H. (2007). A zinc finger-containing glycine-rich RNA- binding protein, atRZ-1a, has a negative impact on seed germination and seedling growth of Arabidopsis thaliana under salt or drought stress conditions, Plant Cell Physiology Vol.48: 1170–1181.

85. Kobayashi, A., Takahashi, A., Kakimoto, Y., Miyazawa, Y., Fujii, N., Higashitani, A. & Takahashi, H. (2007). A gene essential for hydrotropism in roots, Proc. Natl. Acad. Sci. U.S.A. Vol.104: 4724– 4729.

86. Kuhn, J.M. & Schroeder, J.I. (2003). Impacts of altered RNA metabolism on abscisic acid signaling, Current Opinion in Plant Biology Vol.6: 463–469.

87. Lan, L., Li, M., Lai, Y., Xu, W., Kong, Z., Ying, K., Han, B. & Xue, Y. (2005). Microarray analysis reveals similarities and variations in genetic programs controlling pollination/fertilization and

stress responses in rice (Oryza sativa L.), Plant Molecular Biology Vol.59: 151-164.

88. Lareau,L.F., Green, R.E., Bhatnagar, R.S. & Brenner, S.E. (2004). The evolving roles of alternative splicing, Current Opinion in Cell Biology Vol.14: 273–282.
89. Levitt (1972). Responses of plants to environmental stresses. Academic Press, New York.
90. Li, J., Li, X., Guo, L., Lu, F., Feng, X., He, K., Wei, L., Chen, Z., Qu, L. J. & Gu, H. (2006). A subgroup of MYB transcription factor genes undergoes highly conserved alternative splicing in Arabidopsis and rice, Journal of Experimantal Botany Vol. 57: 1263–1273.
91. Li, W.X., Oono, Y., Zhu, J., He, X.J., Wu, J.M., Iida, K., Lu, X.Y., Cui, X., Jin, H. & Zhu, J.K. (2008). The Arabidopsis NFYA5 transcription factor is regulated transcriptionally and posttranscriptionally to promote drought resistance, The Plant Cell Vol.20: 2238–2251.
92. Lilley, J.M. & Kirkegaard, J.A. (2007). Seasonal variation in the value of subsoil water to wheat: Simulation studies in southern New South Wales, Australian Journal of Agricoltural Research Vol.58: 1115–1128.
93. Lin, W.-Y., Matsuoka, D., Sasayama, D. & Nanmori, T. (2010). A splice variant of Arabidopsis mitogen-activated protein kinase and its regulatory function in the MKK6–MPK13 pathway, Plant Science Vol.178: 245–250.
94. Lippold, F., Sanchez, D.H., Musialak, M., Schlereth, A., Scheible, W.R., Hincha, D.K. &, Udvardi, M.K. (2009). AtMyb41 regulates transcriptional and metabolic responses to osmotic stress in Arabidopsis, Plant Physiology Vol.149: 1761–1772.
95. Lira-Medeiros, C.F., Parisod, C., Fernandes, R.A., Mata, C.S., Cardoso, M.A. & Ferreira, P.C. (2010). Epigenetic variation in mangrove plants occurring in contrasting natural environment, PLoS One Vol.5: e10326.
96. Lister, R., O'Malley, R.C., Tonti-Filippini, J., Gregory, B.D., Berry, C.C., et al. (2008). Highly integrated single base resolution maps of the epigenome in Arabidopsis, Cell Vol.133: 523–536.

97. Lorkovic, Z.J. (2009). Role of plant RNA-binding proteins in development, stress response and genome organization, Trends Plant Science Vol.14: 229–236.

98. Lu, X.-Y & Huang, X.-L. (2008). Plant miRNAs and abiotic stress responses, Biochemical and Biophysical Research Communications Vol.368: 458–462.

99. Lynch, J. (1995). Root architecture and plant productivity, Plant Physiology Vol.109: 7–13.

100. Maccaferri, M., Sanguineti, M.C., Corneti, S., Araus Ortega, J.L., Ben Salern, M. et al. (2008). Quantitative trait loci for grain yield and adaptation of durum wheat (Triticum durum Desf.) across a wide range of water availability, Genetics Vol.178: 489-511.

101. Maccaferri, M., Sanguineti, M.C., Giuliani, S. & Tuberosa, R. (2009). Genomics of Tolerance to Abiotic Stress in the Triticeae. In: Feuillet C, Muehlbauer G (eds) Plant Genetics and Genomics: Crops and Models, Vol. 7 Genetics and Genomics of the Triticeae, Springer, pp. 481-558.

102. Manschadi, A.M., Christopher, J., Devoil, P. & Hammer, G.L. (2006). The role of root architectural traits in adaptation of wheat to water-limited environments, Functional Plant Biology Vol.33: 823–837.

103. Masle, J., Gilmore, S.R. & Farquhar, G.D. (2005). The ERECTA gene regulates plant transpiration efficiency in Arabidopsis, Nature Vol.436: 866-870.

104. Mastrangelo, A.M., Belloni, S., Barilli, S., Reperti, B., Di Fonzo, N., Stanca, A.M. & Cattivelli, L. (2005). Low temperature promotes intron retention in two e-cor genes of durum wheat, Planta Vol.221: 705–715.

105. Mastrangelo, A.M., Mazzucotelli, E., Guerra, D., De Vita, P. & Cattivelli, L. (2011a). Improvement of drought resistance in crops: from conventional breeding to genomic selection In "Crop Stress and its Management: Perspectives and Strategies" Arun K. Shanker eds, DOI 10.1007/978-94-007-2220-0_7.

106. Mastrangelo, A.M., Marone, D., Laidò, G., De Leonardis, A.M. & De Vita, P. (2011b). Alternative splicing: Enhancing ability to cope with stress via transcriptome plasticity, Plant Science doi:10.1016/j.plantsci.2011.09.006

107. Matsukura, S., Mizoi, J., Yoshida, T., Todaka, D., Ito, Y., Maruyama, K., Shinozaki, k. & Yamaguchi-Shinozaki, K. (2010). Comprehensive analysis of rice DREB2-type genes that encode transcription factors involved in the expression of abiotic stressresponsive genes, Molecular Genetics & Genomics Vol.283: 185–196.

108. Mazzucotelli, E., Mastrangelo, A.M., Crosatti, C., Guerra, D., Stanca, A.M. & Cattivelli, L. (2008). Abiotic stress response in plants: when post-transcriptional and posttranslational regulations control transcription, Plant Science Vol.174:420–431.

109. McGuire, A.M., Pearson, M.D., Neafsey, D.E. & Galagan, J.E. (2008). Cross-kingdom patterns of alternative splicing and splice recognition, Genome Biology Vol.9: R50.

110. Meuwissen, T., Hayes, B.J. & Goddard, M.E. (2001). Prediction of total genetic value using genome-wide dense marker maps, Genetics Vol.157: 1819–1829.

111. Miller, G., Shulaev, V. & Mittler, R. (2008). Reactive oxygen signaling and abiotic stress, Physiologia Plantarum Vol.133: 481–489.

112. Mirouze, M., Reinders, J., Bucher, E., Nishimura, T., Schneeberger, K., Ossowski, S., Cao, J., Weigel, D., Paszkowski, J. & Mathieu, O. (2009). Selective epigenetic control of retrotransposition in Arabidopsis, Nature Vol.461: 427-430.

113. Miura, K., Jin, J.B. & Hasegawa, P.M. (2007). Sumoylation, a post-translational regulatory process in plants, Current Opinion in Plant Biology Vol.10: 495–502.

114. Mlynarova, L., Nap, J.P. & Bisseling, T. (2007). The SWI/SNF chromatinremodeling gene AtCHR12 mediates temporary growth arrest in Arabidopsis thaliana upon perceiving environmental stress, Plant Journal Vol.51: 874-885.

115. Mohammadi, M., Nat, N.V.K. & Deyholoso, M.K. (2007). Transcriptional profiling of hexaploid wheat (Triticum aestivum L.) roots identifies novel dehydrationresponsive genes, Plant Cell Environ Vol.30: 630-645.

116. Nathan D., Ingvarsdottir K., Sterner D.E., Bylebyl G.R., Dokmanovic M., Dorsey J.A., Whelan K.A., Krsmanovic M., Lane W.S., Meluh P.B. et al. (2006). Histone sumoylation is a

negative regulator in Saccharomyces cerevisiae and shows dynamic interplay with positive-acting histone modifications, Genes Development Vol.20: 966-976.

117. Ner-Gaon, H. & Fluhr, R. (2006). Whole-genome microarray in Arabidopsis facilitates global analysis of retained introns, DNA Research Vol.13: 111–121.

118. Nobuta, K., Venu, R.C., Lu, C., Belo, A., Vemaraju, K., et al. (2007). An expression atlas of rice mRNAs and small RNAs, Nature Biotechnology Vol 25: 473–77.

119. Novillo, F., Alonso, J.M., Ecker, J.R. & Salinas, J. (2004). CBF2/DREB1C is a negative regulator of CBF1/DREB1B and CBF3/DREB1A expression and plays a central role in stress tolerance in Arabidopsis, Proc. Natl. Acad. Sci. U.S.A. Vol.101: 3985–3990.

120. Ogura, Y., Ihara,N., Komatsu, A., Tokioka, Y., Nishioka, M., Takase, T. & Kiyosue, T. (2008).

121. Gene expression, localization, and protein–protein interaction of Arabidopsis SKP1-like (ASK) 20A and 20B, Plant Science Vol.174: 485–495.

122. Oh S.J., Song S.I., Kim Y.S., Jang H.J., Kim M. & Kim Y.K. (2005). Arabidopsis CBF3/DREB1A and ABF3 in transgenic rice increased tolerance to abiotic stress without stunting growth, Plant Physiology Vol.138: 341-351.

123. Oh, S.J., Kwon, C.W., Choi, D.W., Song ,S.I.K. & Kim, J.K. (2007). Expression of barley HvCBF4 enhances tolerance to abiotic stress in transgenic rice, Journal of Plant Biotechnology Vol.5: 646–656.

124. Olsen, A.N., Ernst, H.A., Leggio, L.L. & Skriver, K. (2005). NAC transcription factors: structurally distinct, functionally diverse, Trends in Plant Sciences Vol.10: 79–87.

125. Ooka, H., Satoh, K., Doi, K., Nagata, T., Otomo, Y., Marukami, K., Matsubara, K., Osato, N., Kawai, J., Carninci, P., Hayashizaki, Y., Suzuki, K., Kojima, K., Takahara, Y., Yamamoto, K. & Kikuchi, S. (2003). Comprehensive analysis of NAC family genes in Oryza sativa and Arabidopsis thaliana, DNA Research Vol.10: 239–247.

126. Owttrim, G.W. (2006). RNA helicases and abiotic stress, Nucleic Acids Research Vol. 34(No. 11): 3220-3230.

127. Ozturk, Z.N., Talame, V., Deyholos, M., Michalowski, C.B., Galbraith, D.W., Gozukirmizi, N., Tuberosa, R. & Bohnert, H.J.

(2002). Monitoring large-scale changes in transcript abundance in drought- and salt stressed barley, Plant Molecular Biology Vol.48: 551-573.

128. Palusa, G.S. Ali, G.S. & Reddy, A.S.N. (2007). Alternative splicing of pre-mRNAs of Arabidopsis serine/arginine-rich proteins: regulation by hormones and stresses, Plant Journal Vol.49: 1091–1107.

129. Park, H.Y., Kang, I.S., Han, J.S., Lee, C.H., An, G. & Moon, Y.H. (2009). OsDEG10 encoding a small RNA-binding protein is involved in abiotic stress signaling, Biochemical and Biophysical Research Communication Vol.380: 597–602.

130. Pastori, G.M. & Foyer, C.H. (2002). Common components, networks and pathways of crosstolerance to stress. The central role of 'redox' and abscisic-acid-mediated controls, Plant Physiology Vol.129: 460–468.

131. Peleg, Z., Fahima, T., Krugman, T, Abbo, S., Yakir, D., Korol, A.B. & Saranga, Y. (2009). Genomic dissection of drought resistance in durum wheatxwild emmer wheat recombinant inbreed line population. Plant, Cell and Environment Vol.32: 758–779.

132. Pellegrineschi, A., Reynolds, M., Pacheco, M., Brito, R.M., Almeraya, R., YamaguchiShinozaki, K. & Hoisington, D. (2004). Stress induced expression in wheat of the Arabidopsis thaliana DREB1A gene delays water stress symptoms under greenhouse conditions. Genome Vol.47: 493–500.

133. Petit, J.R., Jouzel, J., Raynaud, D., Barkov, N.I., Barnola, J.-M., Basile, I., Bender, M., Chappellaz, J., Davisk, M. et al. (1999). Climate and atmospheric history of the past 420,000 years from the Vostok ice core, Antarctica, Nature Vol. 399: 429-436.

134. Pinheiro, C. & Chaves, M.M. (2011). Photosynthesis and drought: can we make metabolic connections from available data?, Journal of Experimental Botany Vol.62(No. 3): 869-882.

135. Price, A.H. & Tomos, A.D. (1997). Genetic dissection of root growth in rice (Oryza sativa L.). 2 mapping quantitative trait loci using molecular markers, Theoretical and Applied Genetics Vol.95: 143–152.

136. Qiu, Y. & Yu, D. (2009). Over-expression of the stress-induced OsWRKY45 enhances disease resistance and drought tolerance

in Arabidopsis, Environment and Experimental Botany Vol.65: 35–47.

137. Ramamoorthy, R., Jiang, S.Y., Kumar, N., Venkatesh, P.N. & Ramachandran, S. (2008). A comprehensive transcriptional profiling of the WRKY gene family in rice under various abiotic and phytohormone treatments, Plant Cell Physiology Vol.49: 865–879.

138. Ratcliffe,O.J. & Riechmann, J.L. (2002). Arabidopsis transcription factors and the regulation of flowering time: a genomic perspective, Current Issues in Molecular Biology Vol.4: 77–91.

139. Rebetzke, G.J., Condon, A.G., Richards, R.A. & Farquhar, G.D. (2002). Selection for reduced carbon-isotope discrimination increases aerial biomass and grain yield of rainfed bread wheat, Crop Science Vol. 42: 739-745.

140. Reddy, A.R., Chaitanya, K.V. & Vivekanandan, M. (2004). Drought-induced responses of photosynthesis and antioxidant metabolism in higher plants, Journal of Plant Physiology Vol.161: 1189–1202.

141. Reinders, J., Wulff, B.B.H., Mirouze, M., Marì-Ordònez, A., Dapp, M., Rozhon, W., Bucher, E., Theiler, G. & Paszkowski, J. (2009). Compromised stability of DNA methylation and transposon immobilization in mosaic Arabidopsis epigenomes, Genes Development Vol.23: 939-950.

142. Ren, X., Chen, Z., Liu, Y., Zhang, H., Zhang, M., Liu, Q., Hong, X., Zhu, J.K. & Gong, Z. (2010). ABO3, a WRKY transcription factor, mediates plant responses to abscisic acid and drought tolerance in Arabidopsis, Plant Journal Vol.63: 417–429.

143. Reyes, J.L. & Chua, N.H. (2007). ABA induction of miR159 controls transcript levels of two MYB factors during Arabidopsis seed germination, The Plant Journal Vol.49: 592–606.

144. Rhoades, M.W., Reinhart, B.J., Lim, L.P., Burge, C.P., Bartel, B. & Bartel, D.P. (2002). Prediction of plant microRNA targets, Cell Vol.110: 513-520.

145. Ribaut, J.M. & Ragot, M. (2007). Marker-assisted selection to improve drought adaptation in maize: the backcross approach, perspectives, limitations, and alternatives, Journal of Exp Botany Vol.58: 351-360.

146. Riechmann, J.L., Heard, J., Martin, G., Reuber, L., Jiang, C., Keddie, J., Adam, L., Pineda, O., Ratcliffe, O.J. & Samaha, R.R., Creelman, R., Pilgrim, M., Broun, P., Zhang, J.Z., Ghandehari, D., Sherman, B.K. & Yu, B.K. (2000). Arabidopsis transcription factors: genome-wide comparative analysis among eukaryotes, Science Vol.290: 2105-2110.
147. Rios, G., Gagete, A.P., Castillo. J., Berbel. A., Franco, L. & Rodrigo, M.I. (2007). Abscisic acid and desiccation-dependent expression of a novel putative SNF5-type chromatinremodeling gene in *Pisum sativum*, Plant Physiology and Biochemstry Vol.45: 427-435.
148. Rizhsky, L., Liang, H. & Mittler, R. (2002). The combined effect of drought stress and heat shock on gene expression in tobacco, Plant Physiology Vol.130: 1143–1151.
149. Rizza, F., Ghashghaie, J., Meyer, S., Matteu, L., Mastrangelo, A.M. & Badeck, F.-W. (2011). Constitutive differences in water use efficiency between two durum wheat cultivars, Fields Crops Research doi: 10.1016/J fcr.2011.09.001.
150. Rocak, S. & Linder, P. (2004). DEAD-box proteins: the driving forces behind RNA metabolism. Nature Review Molecular Cell Biology Vol.5: 232–241.
151. Sachetto-Martins, G., Franco, L.O. & deOliveira, D.E. (2000). Plant glycine-rich proteins: a family or just proteins with a common motif?, Biochimica et Biophysica Acta Vol.1492: 1–14.
152. Sade, N., Vinocur, B.J., Diber, A., Shatil, A., Ronen, G., Nissan, H., Wallach, R., Karchi, H. & Moshelion, M. (2009). Improving plant stress tolerance and yield production: is the tonoplast aquaporin SlTIP2; 2 a key to isohydric to anisohydric conversion?, New Phytologist Vol.181: 651–661.
153. Sadras, V.O., & Rodriguez, D. (2007). The limit to wheat water use efficiency in eastern Australia. II. Influence of rainfall patterns, Australian Journal of Agricoltural Research Vol.58: 657–669.
154. Sahi, C., Agarwal, M., Singh, A. & Grover, A. (2007). Molecular characterization of a novel isoform of rice (Oryza sativa L.) glycine rich-RNA binding protein and evidence for its involvement in high temperature stress response, Plant Science Vol.173(No. 2): 144-155.

155. Sanan-Mishra, N., Pham, X.H., Sopory, S.K. & Tuteja,N. (2005). Pea DNA helicase 45 overexpression in tobacco confers high salinity tolerance without affecting yield, Proc. Natl Acad. Sci. USA Vol.102: 509–514.

156. Sarnowski, T.J., Rìos, G,. Jasik, J., Swiezewski, S., Kaczanowski, S., Li, Y., Kwiatkowska, A., Pawlikowska, K., Kozbiał, M., Kozbiał, P. et al. (2005). SWI3 subunits of putative SWI/SNF chromatin-remodeling complexes play distinct roles during Arabidopsis development, Plant Cell Vol.17: 2454-2472.

157. Schoning, J.C., Streitner, C., Meyer, I.M., Gao, Y. & Staiger, D. (2008). Reciprocal regulation of glycine-rich RNA-binding proteins via an interlocked feedback loop coupling alternative splicing to non sense-mediated decay in Arabidopsis, Nucleic Acids Research Vol.36: 6977–6987.

158. Scippa, G.S., Di Michele, M., Onelli, E., Patrignani, G., Chiatante, D. & Bray, E.A. (2004). The histone-like protein H1-S and the response of tomato leaves to water deficit, Journal of Experimental Botany Vol.55: 99-109.

159. Serraj, R. & Sinclair, T.R. (2002). Osmolyte accumulation: can it really increase crop yield under drought conditions?, Plant Cell and Environment Vol.25: 333-341.

160. Shinozahi, Z.K. & Yamaguchi-Shinozaki K. (2000). Molecular responses to dehydration and low temperature: differences and cross-talk between two stress signaling pathways, Current Opinion in Plant Biology Vol. 3: 217–223.

161. Siddique,K.H.M., Tennant, D., Perry, M.W. & Belford, R.K. (1990). Water use and water use efficiency of old and modern wheat cultivars in a Mediterranean-type environment, Australian Journal of Agricultural Research Vol.41(No. 3): 431 – 447.

162. Slafer, G.A., Araus, J.L,, Royo, C. & Del Moral, L.F.G. (2005). Promising ecophysiological traits for genetic improvement of cereal yields in Mediterranean environments, Annals of Applied Biology Vol.146: 61-70.

163. Slotkin, R.K. & Martienssen, R.A. (2007). Transposable elements and the epigenetic regulation of the genome, Nature Reviews Genetics Vol.8: 272-285.

164. Sridha, S. & Wu, K. (2006). Identification of AtHD2C as a novel

regulator of abscisic acid responses in Arabidopsis, Plant Journal Vol.46: 124-133.

165. Sridhar, V.V., Kapoor, A., Zhang, K., Zhu, J., Zhou, T., Hasegawa, P.M.,Bressan, R.A. & Zhu, J.K. (2007). Control of DNA methylation and heterochromatic silencing by histone H2B deubiquitination, Nature Vol.447: 735-738.
166. Steele, K. (2009). Novel upland rice variety bred using marker-assisted selection and clientoriented breeding released in Jharkhand. India: Bangor University.
167. Steele, K.A., Virk, D.S., Kumar, R., Prasad, S.C. & Witcombe, J.R. (2007). Field evaluation of upland rice lines selected for QTLs controlling root traits, Field Crops Research Vol.101: 180–186.
168. Stone, S.L. & Callis, J.(2007). Ubiquitin ligases mediate growth and development by promoting protein death, Current Opinion in Plant Biology Vol.10: 624–632.
169. Sunkar, R. & Zhu, R.-K. (2004). Novel and stress-regulated microRNAs and other small RNAs from Arabidopsis, Plant Cell Vol.16: 2001–2019.
170. Szarzynska, B., Sobkowiak, L., Pant, B.D., Balazadeh, S., Scheible, W., Mueller- Roeber, B., Jarmolowski, A. & Szweykowska-Kulinska, Z. (2009). Gene structures and processing of Arabidopsis thaliana HYL 1-dependent pri-miRNAs, Nucleic Acids Research Vol.37: 3083–3093.
171. Tambussi, E.A., Nogues, S., Ferrio, P., Voltas, J. & Araus, J.L. (2005). Does higher yield potential improve barley performance in Mediterranean conditions?A case of study, Field Crop Research Vol.91: 149-160.
172. Tanner, N.K. & Linder, P. (2001). DExD/H box RNA helicases: from generic motors to specific dissociation functions, Molecular Cell Vol.8: 251–262.
173. Taramino, G., Sauer, M., Stauffer, J.L., Multani, D., Niu, X., Sakai, H. & Hoccholdinger, F. (2007). The maize (Zea mays L.) RTCS gene encodes a LOB domain protein that is a key regulator of embryonic seminal and post-embryonic shoot-borne root initiation, Plant Journal Vol.50: 649–659.
174. Tollenaar, M. & Wu, J. (1999). Yield in temperate maize is attributable to greater stress tolerance, Crop Science Vol.39: 1597-1604.

175. Tollenaar, M. & Lee, E.A. (2002), Yield stability and stress tolerance in maize, Field Crop Research Vol.75: 161-169.

176. Tran, L.S., Nakashima, K., Sakuma, Y., Simpson, S.D., Fujita, Y., Maruyama, K., Fujita, M., Seki, M., Shinozaki, K. & Yamaguchi-Shinozaki, K. (2004). Isolation and functional analysis of Arabidopsis stress inducible NAC transcription factors that bind to a drought responsive cis-element in the early responsive to dehydration stress 1 promoter, Plant Cell Vol.16: 2481–2498.

177. Trujillo, L.E., Sotolongo, M,. Menendez, C., Ochogava, M.E., Coll, Y., Hernandez, I., BorrasHidalgo, O., Thomma, B.P.H.J., Vera, P. & Hernandez, L. (2008). SodERF3, a novel sugarcane ethylene responsive factor (ERF), enhances salt and drought tolerance when over-expressed in tobacco plants, Plant Cell Physiology Vol.49: 512–515.

178. Tuberosa, R., Sanguineti, M.C., Landi, P., Giuliani, M.M., Salvi, S. & Conti,S. (2002a). Identification of QTLs for root characteristics in maize grown in hydroponics and analysis of their overlap with QTLs for grain yield in the field at two water regimes, Plant Molecular Biology Vol.48: 697-712.

179. Tuberosa, R., Salvi, S., Sanguineti, M.C., Landi, P., Maccaferri, M. & Conti, S. (2002b). Mapping QTLs regulating morphophysiological traits and yield: case studies, shortcomings and perspectives in drought stressed maize, Annals of Botany Vol.89: 941–963.

180. Uno, Y., Furihata, T., Abe, H., Yoshida, R., Shinozaki, K. & Yamaguchi-Shinozaki, K. (2000). Arabidopsis basic leucine zipper transcription factors involved in an abscisic aciddependent signal transduction pathway under drought and high-salinity conditions, Proc Natl Acad Sci USA Vol. 97: 11632-11637.

181. Vàgùjfalvi, A., Aprile, A., Miller, A., Dubcovsky, J., Delugu, G., Galiba, G. & Cattivelli, L. (2005). The expression of several Cbf genes at the Fr-A2locus is linked to frost resistance in wheat, Molecular Genetics & Genomics Vol.274: 506–514.

182. van Ginkel, M., Calhoun, D.S., Gebeyehu, G., Miranda, A., Tian-you, C., Pargas Lara, R., Trethowan, R.M., Sayre, K., Crossa, J. & Rajaram, S. (1998). Plant traits related to yield of wheat in early, late, or continuous drought conditions, Euphytica Vol.100: 109-121.

183. Vashisht, A.A., Pradhan, A., Tuteja, R. & Tuteja,N. (2005). Cold- and salinity stress-induced bipolar pea DNA helicase 47 is involved in protein synthesis and stimulated by phosphorylation with protein kinase C, Plant Journal Vol.44: 76–87.

184. Vazquez, F., Gasciolli, V., Crété, P. & Vaucheret, H. (2004). The nuclear dsRNA binding protein HYL1 is required for microRNA accumulation and plant development, but not post-transcriptional transgene silencing, Current Biology Vol.14: 346–351.

185. Verslues, P.E., Agarwal, M., Katiyar-Agarwal, S., Zhu, J. & Zhu, J.K. (2006). Methods and concepts in quantifying resistance to drought, salt and freezing, abiotic stresses that affect plant water status, Plant Journal Vol.45: 523–539.

186. Vicient, C.M., Suoniemi, A., Anamthawat-Jonsson, K., Tanskanen, J., Beharav, A,. et al. (1999). Retrotransposon BARE-1 and its role in genome evolution in the genus Hordeum, Plant Cell Vol.11: 1769-1784.

187. Wang, X.-J., Gaasterland, T. & Chua, N.-H. (2005). Genome-wide prediction and identification of cis-natural antisense transcripts in Arabidopsis thaliana, Genome Biology Vol.6: R30.

188. Wang, B.B. & Brendel, V. (2006). Genome wide comparative analysis of alternative splicing in plants, Proc. Natl. Acad. Sci. USA Vol. 103: 7175–7180.

189. Whitmore, A.P. & Whalley, W.R. (2009). Physical effects of soil drying on roots and crop growth, Journal of Experimental Botany Vol.60: 2845–2857.

190. Xiang, Y., Tang, N,, Du, H., Ye, H. & Xiong, L. (2008). Charaterization of Osb- ZIP23 as a key player of the basic leucine zipper transcription factor family for conferring abscisic acid sensitivity and salinity and drought tolerance in rice, Plant Physiology Vol.148: 1938– 1952.

191. Xiong, L., Gong, Z., Rock, C.D., Subramanian, S., Guo, Y., Xu, W., Galbraith, D. & Zhu, J.K. (2001). Modulation of abscisic acid signal transduction and biosynthesis by an Smlike protein in Arabidopsis, Developmental Cell Vol.1: 771–781.

192. Yang, K.Y., Liu, Y. & Zhang, S. (2001). Activation of a mitogen-activated protein kinase pathway is involved in disease resistance in tobacco, Proc. Natl. Acad. Sci. U.S.A. Vol.98: 741–746.

193. Yang, D.L., Jing, R.L., Chang, X.P. & Li, W. (2007). Identification of quantitative trait loci and environmental interactions for accumulation and remobilization of water-soluble carbohydrates in wheat (Triticum aestivum L.) stems, Genetics Vol.176: 571–584.

194. Yoshida, T., Fujita, Y., Salama, H., Kidokoro, S., Maruyama, K., Mizoi, J., Shinozaki, K. & Yamaguchi-Shinozaki, K. (2010). AREB1, AREB2, and ABF3 are master transcription factors that cooperatively regulate ABRE-dependent ABA signaling involved in drought stress tolerance and require ABA for full activation, Plant Journal Vol.61: 672–685.

195. Zhang, B., Pan, X., Cannon, C.H., Cobb, G.P. & Anderson, T.A. (2006). Conservation and divergence of plant microRNA genes, Plant Journal Vol.46: 243–259.

196. Zhang, K., Sridhar, V.V., Zhu, J., Kapoor, A. & Zhu, J.K. (2007). Distinctive core histone posttranslational modification patterns in Arabidopsis thaliana, PLoS ONE Vol.11: e1210.

197. Zhang, X., Wollenweber, B., Jiang, D., Liu, F. & Zhao, J. (2008). Water deficit and heat shock effects on photosynthesis of a transgenic Arabidopsis thaliana constitutively expressing ABP9, a bZIP transcription factor, Journal of Experimental Botany Vol. 59: 839–848.

198. Zhang, G., Chen, M., Li, L., Xu, Z., Chen, X., Guo, J. & Ma, Y. (2009). Overexpression of the soybean Gm*E*RF3 gene, an AP2/ERF type transcription factor for increased tolerances to salt, drought, and diseases in transgenic tobacco, Journal of Experimental Botany Vol.60 (No. 13): 3781–3796.

199. Zhao, B.T., Liang, R.Q., Ge, L.F., Li, W., Xiao, H.S., Lin, H.X., Ruan, K.C. & Jin, Y.X. (2007). Identification of drought-induced microRNAs in rice, Biochemical and Biophysical Research Communication Vol.354: 585–590.

200. Zhao, B., Ge, L., Liang, R., Li, W., Ruan, K., Lin, H. & Jin, Y. (2009). Members of miR-169 family are induced by high salinity and transiently inhibit the NF-YA transcription factor, BMC Molecular Biology Vol.10: 29.

201. Zheng, X., Chen, B., Lu, G. & Han, B. (2009). Overexpression of a NAC transcription factor enhances rice drought and salt tolerance,

Biochemical and Biophysical Research Communication Vol.379: 985–989.

202. Zhu, Q., Zhang, J., Gao, X., Tong, J., Xiao, L., Li, W. & Zhang, H. (2010). The Arabidopsis AP2/ERF transcription factor RAP2.6 participates in ABA, salt and osmotic stress responses, Gene Vol.457: 1–12.

203. Zoon, F.C. &. Van, Tienderen P.H. (1990). A rapid quantitative measurement of root length and root branching by microcomputer image analysis, Plant Soil Vol.126: 301-308.

Chapter 3

Plant Physiology Meets Phytopathology: Plant Primary Metabolism and Plant–Pathogen Interactions

Susanne Berger[1], Alok K. Sinha[2], and Thomas Roitsch[1]

[1]Julius-von-Sachs-Institut fuer Biowissenschaften, Universitaet Wuerzburg, Julius-von-Sachs-Platz 2, 97082 Wuerzburg, Germany

[2]National Institute for Plant Genome Research, Aruna Asaf Ali Road, New Delhi 110067, India

ABSTRACT

Phytopathogen infection leads to changes in secondary metabolism based on the induction of defence programmes as well as to changes in primary metabolism which affect growth and development of the plant. Therefore, pathogen attack causes crop yield losses even in interactions which do not end up with disease or death of the plant. While the regulation of defence responses has been intensively studied for decades, less is known about the effects of pathogen infection on primary metabolism. Recently, interest in this research area has been growing, and aspects of photosynthesis, assimilate partitioning, and source–sink regulation in different types of plant–pathogen interactions have been investigated. Similarly, phytopathological studies take into consideration the physiological status of the infected tissues to elucidate the fine-tuned infection mechanisms. The aim of this review is to give a summary of recent advances in the mutual interrelation between primary metabolism and pathogen infection, as well as to indicate current developments in non-invasive techniques and important strategies of combining modern molecular and physiological techniques with phytopathology for future investigations.

PLANT–PATHOGEN INTERACTIONS

Plant pathogens include fungi, bacteria, oomycetes, and viruses. Pathogens have devised different strategies to invade a plant, as well as to feed on and reproduce in the plant. Besides the assignment to bacteria or fungi, this is regarded as an important feature to classify the attacking micro-organism (Oliver and Ipcho, 2004). Biotrophic pathogens need living tissue for growth and reproduction; in many interactions the tissue will die in the late stages of the infection (hemi-biotrophic pathogens). By contrast, necrotrophic pathogens kill the host tissue at the beginning of the infection and feed on the dead tissue. Viruses, in general, need living tissue for nutrition, while biotrophic as well as necrotrophic strategies can be found among bacteria and fungi. Similarities in the pathways involved in the defence of the plants against biotrophic fungi and bacteria on one hand or against necrotrophic fungi and bacteria on the other hand have been described. The jasmonate/ethylene pathway is more important in defending necrotrophic pathogens while salicylic acid-dependent responses are more effective against biotrophic pathogens (Thomma *et al.*, 2001).

Pathogens can also be divided according to the environment in which they occur and the tissues which they infect. A common classification is to distinguish between above- and below-ground tissues as the primary target of the pathogen. Related to this, above-ground tissue might be green, assimilate-producing tissue, typically source leaves, or assimilate-importing tissue such as flowers. Pathogens infecting source tissue will encounter different conditions related to primary metabolism as well as to defence responses compared with those pathogens infecting sink or assimilate-producing tissue such as roots, flowers, and sink leaves. The investigation and the understanding of interactions with source-(leaf)-pathogens is more advanced and, therefore, this review will focus on these interactions.

Plants are resistant to the majority of micro-organisms based on preformed and induced defence mechanisms. Recognition of the presence of micro-organisms is the first step for the activation of defence responses. Basal resistance begins with the recognition of elicitors which are derived from the micro-organisms, and similarities in the recognition of pathogen-associated molecular patterns (PAMPs) in animals and in plants have been documented (Nuernberger and Scheel, 2001;Nuernberger and Lipka, 2005). Upon contact with pathogens or with non-pathogenic micro-organisms or elicitors, ion fluxes, phosphorylation/dephosphorylation of proteins, and the production of signalling molecules such as salicylic acid, jasmonic acid, ethylene, and reactive oxygen species are activated. This leads to the regulation of gene expression and the induction of defence responses, for example, cell wall strengthening and the accumulation of phytoalexins and pathogenesis related (PR) proteins (Dangl and Jones, 2001; Garcia-Brugger *et al.*, 2006). According to the current model, some micro-organisms became virulent by the production of effector molecules which contribute to their virulence, for example, by the suppression of plant defence (Jones and Dangl, 2006). In these compatible interactions the virulent pathogen can spread in the susceptible plant.

Specific resistance is based on the recognition of the activity of these effector molecules by plant receptor proteins. In these incompatible interactions the plant is resistant and can successfully prevent the pathogen spreading. The successful defence is based on the early recognition of avirulent strains of plant pathogens and the fast activation of defence (Jones and Dangl, 2006). Furthermore, the recognition of the avirulent strains activates, in addition to the already mentioned defence reactions, a localized programmed cell death which can efficiently halt the spreading of biotrophic pathogens (Heath, 2000).

EFFECTS ON PHOTOSYNTHESIS, SUGAR ACCUMULATION, AND SINK METABOLISM

Photosynthesis

There are obvious reasons why a contact with pathogens alters plant primary metabolism. It has been shown that the induction of defence is cost-intensive (Heil and Bostock, 2002; Swarbrick *et al.*, 2006). This causes an increased demand for assimilates in the plant. In addition, the pathogen tries to manipulate plant carbohydrate metabolism for its own need, which is, for example, evident in the production of opines in plants infected with *Agrobacterium tumefaciens*. The withdrawal of nutrients by the pathogen will further increase the demand for assimilates. Furthermore, pathogen infection often leads to the development of chlorotic and necrotic areas and to a decrease in photosynthetic assimilate production. The effect on photosynthesis can be analysed by monitoring*in vivo* chlorophyll fluorescence. This method is based on measuring the fluorescence of chlorophyll *a* in a dark-adapted plant and after saturating light pulses (Schreiber *et al.*, 1986; Schreiber, 2004). This fluorescence is a very sensitive marker for the efficiency of photosynthesis since the energy of absorbed photons can either be used for photosynthetic electron transport or the energy will be dissipated as heat or fluorescence (van Kooten and Snel, 1990). Therefore, chlorophyll fluorescence responds to the changes in energy conversion at photosystem II reaction centres and is also sensitive to any limitations in the dark enzymatic steps of the complex process of photosynthesis (Govindjee, 2004). Analysis of chlorophyll fluorescence is non-invasive and therefore time-courses can be performed on the same plant material.

Using this method, down-regulation of effective photosystem II quantum yield in compatible interactions with biotrophic as well as necrotrophic pathogens has been reported. This comprises interactions with biotrophic bacteria such as *Pseudomonas syringae* (Bonfig *et al.*, 2006), biotrophic fungi such as *Albugo candida* (Chou *et al.*, 2000),

Puccinia coronata and*Blumeria graminis* (Scholes and Rolfe, 1996; Swarbrick *et al.*, 2006), as well as necrotrophic fungi such as *Botrytis cinerea* (Berger *et al.*, 2004) and viruses such as tobacco mosaic virus and abutilon mosaic virus (Balachandran *et al.*, 1994; Lohaus *et al.*, 2000; Perez-Bueno *et al.*, 2006). In several cases, changes in chlorophyll fluorescence were detectable earlier than symptoms were visible by eye demonstrating the sensitivity of this technique.

A substantial advancement of chlorophyll fluorescence analysis is chlorophyll fluorescence imaging (Oxborough, 2004, and references therein). In addition to monitoring the temporal dynamics, this method enables the spatial resolution of photosynthesis. Figure 1 illustrates an example of the application of chlorophyll fluorescence imaging to an*Arabidopsis* leaf infected with two different pathogens, *P. syringae* on one half and *B. cinerea* on the other half. Using chlorophyll fluorescence imaging, it was reported that, in general, the changes in photosynthesis upon infection are local. In addition, the imaging technology revealed the complexity and heterogeneity of effects. In *Arabidopsis* leaves infected with *A. candida* and in tomato plants infected with *B. cinerea* a ring of enhanced photosynthesis was detectable surrounding the area with decreased photosynthesis at the infection site. At present it is not clear if this stimulation of photosynthesis is due to the defence strategy of the plant.

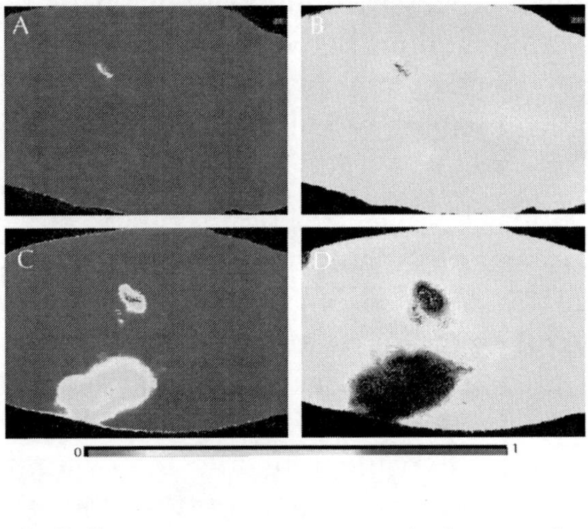

Figure 1: Chlorophyll fluorescence imaging of plant–pathogen interactions. A leaf of *Arabidopsis* was infected with *B. cinerea* on the upper half of the leaf and 24 h later with a virulent strain of *P. syringae* on the lower half of the

leaf. Measurements were performed 6 h and 48 h after the *P. syringae*infection. Shown are false colour images of the maximum PSII quantum yield (A, C) and effective PSII quantum yield (B, D). The first time point corresponding to 6 h after inoculation with *P. syringae* and 30 h after inoculation with *B. cinerea* is shown in (A) and (B). The second time point corresponding to 48 h after inoculation with *P. syringae* and 72 h after inoculation with *B. cinerea* is shown in (C) and (D).

A decrease in photosynthesis has also been reported in incompatible interactions (Scharte *et al.*, 2005; Bonfig *et al.*, 2006; Swarbrick *et al.*, 2006). A direct comparison of the interaction of *Arabidopsis* with a virulent and an avirulent strain of *P. syringae* showed that the major difference in the changes in photosynthesis was the speed of the effects. A decrease in photosynthesis was detectable earlier with the avirulent strain than with the virulent strain. This indicates a similarity between the effects on primary and secondary metabolism since, in both cases, reactions are earlier in the incompatible interaction while qualitatively the responses in both interactions are similar (Tao *et al.*, 2003). Chlorophyll fluorescence imaging of tobacco leaves in a macro- and microscopic scale during an incompatible interaction with *Phytophthora nicotianae* revealed that the decline in photosynthesis is a highly localized process which occurs in single mesophyll cells (Scharte *et al.*, 2005). The results led to the proposal that plants switch off photosynthesis and other assimilatory metabolism to initiate respiration and other processes required for defence. Also in compatible and incompatible interactions between barley and *B. graminis* causing powdery mildew, photosynthesis was decreased (Swarbrick *et al.*, 2006). Quantitative imaging of chlorophyll fluorescence showed that photosynthesis was reduced both in cells directly below the fungal colonies and in adjacent cells during the compatible interaction.

Expression of sugar-regulated photosynthetic genes, such as the small subunit of ribulose-1,5-*bis*phosphate carboxylase (*RbcS*) and chlorophyll*a,b* binding protein (*Cab*) was analysed and, in most cases, in agreement with the results of chlorophyll fluorescence analysis, a down-regulation of the expression after pathogen infection was reported. However, in two incompatible interactions, no repression of photosynthetic genes was detectable even though photosynthetic activity decreased (Bonfig *et al.*, 2006; Swarbrick *et al.*, 2006). This indicates that a decrease in photosynthetic activity is not necessarily preceded by the repression of photosynthetic genes.

Sink Metabolism and Sugar Accumulation

The down-regulation of photosynthesis and the simultaneous increased demand for assimilates very often leads to a transition of source tissue into sink tissue during plant–pathogen interactions. One indication for the induction of a sink status in infected leaves is the increase of cell wall invertase activity. Cell wall invertases are extracellular enzymes which cleave sucrose in the apoplast into glucose and fructose. The resulting hexoses are transported by hexose transporters into the cell. Therefore, extracellular invertases are important for apoplastic phloem unloading and key enzymes in determining sink strength (Roitsch *et al.*, 2003). The cleavage of extracellular sucrose will also result in the decreased export of assimilates from the tissue. Enhanced expression and activity of cell wall invertases has been reported in several plant–pathogen interactions (Benhamou *et al.*, 1991; Chou *et al.*, 2000; Fotopoulos *et al.*, 2003; Berger*et al.*, 2004; Bonfig *et al.*, 2006; Swarbrick *et al.*, 2006). Similarly, reduced sucrose export from infected source leaves has been observed (Scharte *et al.*, 2005). Even though a role for extracellular invertases in plant–pathogen interaction has been suggested in all of these studies, details of the causal relationship between this increase in invertase activity and the outcome of the interaction are still not clear (see below).

Even though the repression of photosynthesis and the induction of sink metabolism seem to be a general response to pathogen infection, the effect on sugar levels varies considerably between different plant–pathogen interactions. Infection of tobacco with tobacco mosaic virus or *P. nicotianae*, of wheat with *Puccinia graminis* and of *Arabidopsis* with *A. candida* results in an increase in the levels of soluble sugars (Wright *et al.*, 1995; Chou *et al.*, 2000; Herbers *et al.*, 2000; Scharte *et al.*, 2005). By contrast, sugar levels in *Arabidopsis* are not altered by infection with *P. syringae* and decrease in tomato plants after inoculation with *B. cinerea*(Berger *et al.*, 2004; Bonfig *et al.*, 2006) as well as in sunflowers treated with *Sclerotinia sclerotiorum* (Jobic *et al.*, 2007). In the tomato–grey mould interaction, levels of sucrose decrease more than the levels of hexoses, leading to an increase in the hexose-to-sucrose ratio. This increase in the hexose-to-sucrose ratio could be due to the enhanced activity of invertases. For many of the studies, sugar levels have been determined from a region containing a large area of and around the infection site. The analysis of sugar levels and invertase activity in infected versus uninfected regions of an inoculated leaf showed only strong effects in the infected region (Chou *et al.*, 2000; Swarbrick *et al.*, 2006). This strongly supports the importance of spatial resolution and indicates that most measurements missed or underestimated changes. In addition, it would be desirable to distinguish between intracellular and apoplastic sugar levels, and protocols for the isolation of apoplastic fluid have been described (Lohaus *et al.*, 2001). Several pathogens live in the apoplast, therefore, extracellular assimilate levels might be more relevant for the pathogen than intracellular levels. Scharte *et al.* (2005) reported an increase in the levels of apoplastic sucrose and hexose levels as well as invertase activity in tobacco after infection with *P. nicotianae*. The use of methods providing information on the spatial distribution of compounds on the organ-, tissue-, and subcellular level is necessary to elucidate the effects. Sugars are not only nutrients needed for growth, respiration, and the accumulation of storage compounds, but, in addition, sugars are signals which can regulate gene expression (Koch, 1996). For instance, the down-regulation of photosynthetic genes has been attributed to the accumulation of hexose sugars (Scholes *et al.*, 1994; Chou *et al.*, 2000;Pego *et al.*, 2000; Berger *et al.*, 2004). Using photoautotrophic cell cultures of tomato or *Chenopodium rubrum*, it was shown that treatment with fungal elicitors or glucose resulted in a rapid decrease in photosynthesis and later in the down-regulation of photosynthetic gene expression and the induction of cell wall invertase expression (Ehness *et al.*, 1997; Sinha *et al.*, 2002). Similarly to the response to avirulent strains mentioned above, a decrease in net photosynthesis is also not preceded by the repression of photosynthetic genes in these systems. These data strongly suggest that the reduction in the rate of photosynthesis is a fast and immediate effect of pathogen infection, while down-regulation of photosynthetic gene expression is a slower process.

Models for the regulation of carbohydrate metabolism and photosynthesis by elicitors or pathogens have been proposed earlier (Roitsch, 2004;Walters and McRoberts, 2006). Figure 2 presents a model for the most investigated type of interaction, the infection with virulent biotrophic source pathogens. Pathogen attack first initiates a series of rapid changes resulting in a decline in photosynthesis and an increase in respiration, photorespiration, and invertase enzyme activity. The mechanisms and pathways which mediate these rapid changes are largely unknown. The electrophilic oxylipin 12-oxo-phytodienoic acid is a compound which has been shown to accumulate after pathogen infection and to result in a decrease in photosynthesis very shortly after application, suggesting that it might be involved in the decrease in photosynthesis upon pathogen challenge (Berger *et al.*, 2007). Hexoses released by the action of increased invertase activity act as signalling molecules and repress photosynthetic genes. This down-regulation of photosynthetic genes, in turn, again decreases the net photosynthesis rate (Fig. 2). While the data from several plant–pathogen interactions, especially with virulent biotrophic fungal pathogens, fit into this general model, there are also some examples that differ in distinct points from this model. As discussed above, the accumulation of hexoses and the repression of photosynthetic genes have not always been observed. Another example is that the expression, but not the activity, of cell wall invertases is increased in the *Arabidopsis–P. syringae* interaction. These exceptions from the rule support the complexity of the interactions which is based on the fundamental diversity of the plant as well as the microbial partner. As discussed above, in this context it has to be taken into account that the analysis of gene expression and sugar levels typically integrates over a bigger area so that local effects observed by imaging techniques might not be detectable.

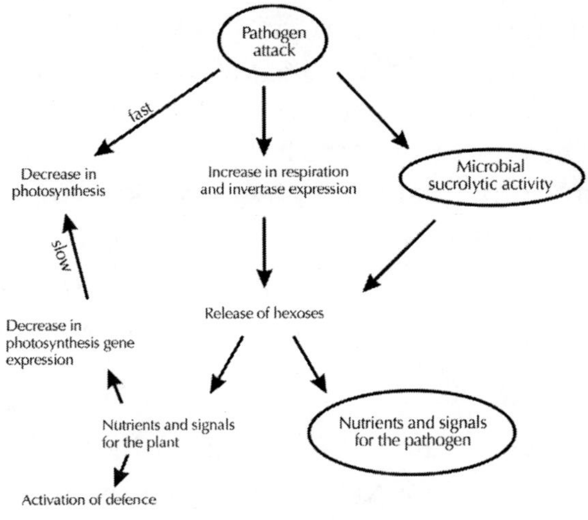

Figure 2: Model of changes in carbohydrate metabolism in response to infection with virulent, biotrophic pathogens. Components concerning the microbial partner are encircled.

RELEVANCE OF THE REGULATION OF CARBOHYDRATE METABOLISM FOR PLANT–PATHOGEN INTERACTIONS

Even though several reports describe the effect of pathogen infection on carbohydrate metabolism, there is still a considerable lack of knowledge regarding how these changes influence the outcome of plant–pathogen interactions. There are several factors contributing to the complexity of the mutual relation between carbohydrate status and development of disease/resistance. First, as discussed above, the carbohydrate status affects the defence as well as general metabolism of the plant. Second, sugars are not only nutrients and signals for the plant partner but also for the microbial partner. Therefore, changes in assimilate levels may influence the spreading of the pathogen and might regulate gene expression of the pathogen. Third, certain pathogens also possess extracellular sucrolytic enzymes such as invertases, fructoexohydrolases, and levansucrases. The presence of extracellular invertases has been observed in *B. cinerea* and *Uromyces fabae* (Geissmann *et al.*, 1991;Voegele *et al.*, 2006) and suggested for *A. candida* (Chou *et al.*, 2000). With expression of these enzymes the pathogen would be able to alter the hexose and sucrose levels in the apoplast (Fig. 2). It still needs to be investigated if these microbial sucrolytic activities are important for pathogenicity. This implies that the investigation of the plant side needs to be complemented by the analysis of the microbial part.

Functional approaches are necessary to elucidate how alterations in carbohydrate metabolism affect disease development and resistance induction. In one transgenic approach, tobacco plants constitutively expressing a yeast invertase were generated (von Schaewen *et al.*, 1990). Expression of the yeast-derived invertase in the cell wall of tobacco and *Arabidopsis* plants leads to the accumulation of carbohydrates and the inhibition of photosynthesis and strongly influences growth. These plants also showed increased defence gene expression and enhanced resistance against tobacco mosaic virus (Herbers *et al.*, 1996). This is in agreement with the induction of defence reactions by hexoses generated by increased invertase activity. Another functional approach to study the role of invertases is the analysis of knock-out or antisense plants. *Arabidopsis* plants containing a single knock-out in each of the four cell wall invertase genes did not show an obvious alteration in resistance against *P. syringae*or *A. brassicicola* (C Bandulet, S Berger, T Roitsch, unpublished observation). This lack of clear effects might be due to redundancy. One strategy to overcome this problem is the modulation of invertase activity on a post-translational level. Proteinaceous inhibitors of invertases are produced by higher plants. Down-regulation of invertase activity by these plant invertase inhibitors could be used to study the function of invertases. This approach has already been successfully applied to investigate their role in senescence (Balibrea Lara *et al.*, 2004). Other post-translational regulatory mechanisms are currently not known (Huang*et al.*, 2007).

SUMMARY AND FUTURE PERSPECTIVES

The down-regulation of photosynthesis and the induction of sink metabolism have been regarded as general responses of source tissue to infection. As described above, the comparison of the changes induced by different pathogens revealed the complexity and divergence of the responses. Therefore, careful investigations of different aspects of carbohydrate metabolism in each pathosystem are required to understand the basis of the interaction in more detail. While the changes caused by infection with virulent biotrophic fungi are the best understood, research is needed to elucidate the interaction with biotrophic bacteria, viruses, and necrotrophic pathogens as well as avirulent micro-organisms. This applies also to interactions in which sink organs such as roots, flowers, and stems are the first targets. Source tissue, as an assimilate exporting tissue, has a completely different carbohydrate status and enzymatic constitution than sink tissue. The data obtained with source pathogens would lead to the expectation that, due to the increased demand for assimilates, infection of sink tissue leads to a further increase in sink metabolism. In agreement with this hypothesis, extracellular invertase accumulated in tomato roots infected with *Fusarium oxysporum*(Benhamou *et al.*, 1991) and expression of sucrose synthase and starch synthase was induced in roots infected with *Plasmodiophora brassicae*(Siemens *et al.*, 2006). Tumours induced by *Agrobacterium tumefaciens* in *Arabidopsis* also showed increased expresssion of several sucrose-degrading enzymes (Deeken *et al.*, 2006). In addition, data obtained from symbiotic interactions also indicate regulation of carbohydrate fluxes; one example is the induction of invertase and sucrose synthase in mycorrhizal arbuscules (Blee and Anderson, 2002; Schaarschmidt *et al.*, 2006). Besides the effects on carbohydrate metabolism, the molecular mechanisms of regulation will be interesting to elucidate. As discussed above, one of the important and basic mechanisms generally believed to contribute to up-regulation of defences and down-regulation of photosynthesis in the interaction of pathogens with source tissue are the increases in hexose levels or the increase in the hexose-to-sucrose ratio. In sink tissue, this ratio is already very high but, nevertheless, infection leads to further stimulation of sink metabolism. In addition, despite high hexose levels, sink tissues do not exhibit constitutively activated defence and general resistance. This indicates that a combination of metabolic signals such as hexoses and phytohormone signals is responsible for regulation of carbohydrate and defence metabolism.

There are several technical future directions which need to be pursued in order to elucidate the connection between pathogen infection and plant primary metabolism. Methods for spatial analysis of metabolites, for example, sucrose, glucose, and ATP at the single cell dimension have been reported (Mueller-Klieser and Walenta, 1993; Borisjuk *et al.*, 2002). The application of the non-invasive method of chlorophyll fluorescence imaging has revealed the importance of analysing spatio-temporal changes. The use of different measuring protocols combined with non-biased approaches for data analysis (Matous *et al.*, 2006) should improve the identification of pathogen-induced fluorescence signatures. Imaging a combination of processes such as gene expression, assimilate and secondary metabolites levels, and microbial spreading should further enrich stress physiological studies. Non-invasive techniques such as fluorescence-labelled pathogens, reporter gene expression, ^{11}C-labelled assimilates (Schwachtje *et al.*, 2006), FRET sugar nanosensors (Deuschle*et al.*, 2006), multicolour fluorescence imaging (Chaerle *et al.*, 2007; Lenk*et al.*, 2007), and spontaneous photon emission (Bennett *et al.*, 2005) are available.

As already pointed out, functional approaches to modulate carbohydrate metabolism and signalling are required to investigate how the carbohydrate status and its regulation influence plant–pathogen interactions.

Most research focuses, for obvious reasons, either on the plant side or on the pathogen side. A combination of investigations of both partners including modern imaging technology and functional approaches is of central importance to understand the molecular and physiological basis for plant–pathogen interactions. This knowledge will also support the development of strategies to increase pathogen resistance in plants for practical applications.

ACKNOWLEDGMENTS

We apologize to the colleagues whose work was not cited due to space limitations. This work was supported by Bayerisches Staatsministerium für Umwelt, Gesundheit und Verbraucherschutz and the SFB 567.

REFERENCES

1. Balachandran S, Osmond CB, Daley PF. 1994. Diagnosis of the earliest strain-specific interactions between tobacco mosaic virus and chloroplasts of tobacco leaves in vivo by means of chlorophyll fluorescence imaging. Plant Physiology 104, 1059–1065.
2. Balibrea Lara ME, Gonzalez Garcia MC, Fatima T, Ehness R, Lee TK, Proels R, Tanner W, Roitsch T. 2004. Extracellular invertase is an essential component of cytokinin-mediated delay of senescence. The Plant Cell 16, 1276–1287.
3. Benhamou N, Grenier J, Chrispeels MJ. 1991. Accumulation of beta-fructosidase in the cell walls of tomato roots following infection by a fungal wilt pathogen. Plant Physiology 97, 739–750.
4. Bennett M, Mehta M, Grant M. 2005. Biophoton imaging: a nondestructive method for assaying R gene responses. Molecular Plant–Microbe Interactions 18, 95–102.
5. Berger S, Benediktyova Z, Matous K, Bonfig KB, Mueller MJ, Nedbal L, Roitsch T. 2007. Visualization of dynamics of plant–pathogen interaction by novel combination of chlorophyll

fluorescence imaging and statistical analysis: differential effects of virulent and avirulent strains of P. syringae and of oxylipins on A. thaliana. Journal of Experimental Botany 58, 797–806.

6. Berger S, Papadopoulos M, Schreiber U, Kaiser W, Roitsch T. 2004. Complex regulation of gene expression, photosynthesis and sugar levels by pathogen infection in tomato. Physiologia Plantarum 122, 419–428.

7. Blee KA, Anderson AJ. 2002. Transcripts for genes encoding soluble acid invertase and sucrose synthase accumulate in root tip and cortical cells containing mycorrhizal arbuscules. Plant Molecular Biology 50, 197–211.

8. Bonfig KB, Schreiber U, Gabler A, Roitsch T, Berger S. 2006. Infection with virulent and avirulent P. syringae strains differentially affects photosynthesis and sink metabolism in Arabidopsis leaves. Planta 225, 1–12.

9. Borisjuk L, Walenta S, Rolletschek H, Mueller-Klieser W, Wobus U, Weber H. 2002. Spatial analysis of plant metabolism: sucrose imaging within Vicia faba cotyledons reveals specific developmental patterns. The Plant Journal 29, 521–530.

10. Chaerle L, Lenk S, Hagenbeek D, Buschmann C, Van Der Straeten D. 2007. Multicolor fluorescence imaging for early detection of the hypersensitive reaction to tobacco mosaic virus. Journal of Plant Physiology 164, 253–262.

11. Chou H, Bundock N, Rolfe S, Scholes J. 2000. Infection of Arabidopsis thaliana leaves with Albugo candida causes a reprogramming of host metabolism. Molecular Plant Pathology 1, 99–113.

12. Dangl JL, Jones JD. 2001. Plant pathogens and integrated defence responses to infection. Nature 411, 826–833.

13. Deeken R, Engelmann JC, Efetova M, et al. 2006. An integrated view of gene expression and solute profiles of Arabidopsis tumors: a genome-wide approach. The Plant Cell 18, 3617–3634.

14. Deuschle K, Chaudhuri B, Okumoto S, Lager I, Lalonde S, Frommer WB. 2006. Rapid metabolism of glucose detected with FRET glucose nanosensors in epidermal cells and intact roots of Arabidopsis RNA-silencing mutants. The Plant Cell 18, 2314–2325.

15. Ehness R, Ecker M, Godt DE, Roitsch T. 1997. Glucose and stress independently regulate source and sink metabolism and defense mechanisms via signal transduction pathways involving protein phosphorylation. The Plant Cell 9, 1825–1841.
16. Fotopoulos V, Gilbert MJ, Pittman JK, Marvier AC, Buchanan AJ, Sauer N, Hall JL, Williams LE. 2003. The monosaccharide transporter gene, AtSTP4, and the cell-wall invertase, Atbetafruct1, are induced in Arabidopsis during infection with the fungal biotroph Erysiphe cichoracearum. Plant Physiology 132, 821–829.
17. Garcia-Brugger A, Lamotte O, Vandelle E, Bourque S, Lecourieux D, Poinssot B, Wendehenne D, Pugin A. 2006. Early signaling events induced by elicitors of plant defenses. Molecular Plant–Microbe Interactions 19, 711–724.
18. Geissmann M, Frey T, Ruffner H. 1991. Occurence and properties of acidic invertases in cultures of B. cinerea. Mycologia Research 95, 1321–1327.
19. Govindjee. 2004. Chlorophyll a fluorescencea a bit of basics and history. In: Papageorgiu G, Govindjee, eds. Chlorophyll a fluorescence: a signature of photosynthesis. Dordrecht, The Netherlands: Springer, 1–42. Heath MC. 2000. Hypersensitive response-related death. Plant Molecular Biology 44, 321–334.
20. Heil M, Bostock RM. 2002. Induced systemic resistance (ISR) against pathogens in the context of induced plant defences. Annals of Botany 89, 503–512.
21. Herbers K, Meuwly P, Frommer WB, Metraux JP, Sonnewald U. 1996. Systemic acquired resistance mediated by the ectopic expression of invertase: possible hexose sensing in the secretory pathway. The Plant Cell 8, 793–803.
22. Herbers K, Takahata Y, Melzer M, Mock HP, Hajirezaei M, Sonnewald U. 2000. Regulation of carbohydrate partitioning during the interaction of potato virus Y with tobacco. Molecular Plant Pathology 1, 51–59.
23. Huang LF, Bocock P, Davis J, Koch K. 2007. Regulation of invertase: a suite of transcriptional and post-transcriptional mechanisms. Functional Plant Biology 34, 499–507.

24. Jobic C, Boisson AM, Gout E, Rascle C, Fevre M, Cotton P, Bligny R. 2007. Metabolic processes and carbon nutrient exchanges between host and pathogen sustain the disease development during sunflower infection by Sclerotinia sclerotiorum. Planta 226, 251–265.
25. Jones JD, Dangl JL. 2006. The plant immune system. Nature 444, 323–329.
26. Koch KE. 1996. Carbohydrate-modulated gene expression in plants. Annual Review of Plant Physiology and Plant Molecular Biology 47, 509–540.
27. Lenk S, Chaerle L, Pfundel EE, Langsdorf G, Hagenbeek D, Lichtenthaler HK, Van Der Straeten D, Buschmann C. 2007. Multispectral fluorescence and reflectance imaging at the leaf level and its possible applications. Journal of Experimental Botany 58, 807–814.
28. Lohaus G, Pennewiss K, Sattelmacher B, Hussmann M, Muehling KH. 2001. Is the infiltration-centrifugation technique appropriate for the isolation of apoplastic fluid? A critical evaluation with different plant species. Physiologia Plantarum 111, 457–465.
29. Lohaus G, Heldt HW, Osmond CB. 2000. Infection with phloem limited Abutilon mosaic virus causes localized carbohydrate accumulation in leaves of Abutilon striatum: relationships to symptom development and effects on chlorophyll fluorescence quenching during photosynthetic induction. Plant Biology 2, 161–167.
30. Matous K, Benediktyova Z, Berger S, Roitsch T, Nedbal L. 2006. Case study of combinatorial imaging: what protocol and what chlorophyll fluorescence image to use when visualizing infection of Arabidopsis thaliana by Pseudomonas syringae? Photosynthesis Research 90, 243–253.
31. Mueller-Klieser W, Walenta S. 1993. Geographical mapping of metabolites in biological tissue with quantitative bioluminescence and single photon imaging. Histochemical Journal 25, 407–420.
32. Nuernberger T, Lipka V. 2005. Non-host resistance in plants: new insights into an old phenomenon. Molecular Plant Pathology 6, 1–11.

33. Nuernberger T, Scheel D. 2001. Signal transmission in the plant immune response. Trends in Plant Science 6, 372–379.
34. Oliver RP, Ipcho SVS. 2004. Arabidopsis pathology breathes new life into the necrotrophs-vs.-biotrophs classification of fungal pathogens. Molecular Plant Pathology 5, 347–352.
35. Oxborough K. 2004. Imaging of chlorophyll fluorescence: theoretical and practical aspects of an emerging technique for the monitoring of photosynthetic performance. Journal of Experimental Botany 55, 1195–1205.
36. Pego JV, Kortstee AJ, Huijser C, Smeekens SC. 2000. Photosynthesis, sugars and the regulation of gene expression. Journal of Experimental Botany 51, 407–416.
37. Perez-Bueno ML, Ciscato M, VandeVen M, Garcia-Luque I, Valcke R, Baron M. 2006. Imaging viral infection: studies on Nicotiana benthamiana plants infected with the pepper mild mottle tobamovirus. Photosynthesis Research 90, 111–123.
38. Roitsch T. 2004. Plant response to stress: source–sink regulation by stress. In: Goodman RM, ed. Encyclopedia of plant and crop science. New York: Marcel Dekker Inc, 1010–1013.
39. Roitsch T, Balibrea ME, Hofmann M, Proels R, Sinha AK. 2003. Extracellular invertase: key metabolic enzyme and PR protein. Journal of Experimental Botany 54, 513–524.
40. Schaarschmidt S, Roitsch T, Hause B. 2006. Arbuscular mycorrhiza induces gene expression of the apoplastic invertase LIN6 in tomato (Lycopersicon esculentum) roots. Journal of Experimental Botany 57, 4015–4023.
41. Scharte J, Scho¨n H, Weis E. 2005. Photosynthesis and carbohydrate metabolism in tobacco leaves during an incompatible interaction with Phytophthora nicotianae. Plant, Cell and Environment 28, 1421–1435.
42. Scholes J, Rolfe SA. 1996. Photosynthesis in localised regions of oat leaves infected with crown rust (Puccinia coronata): quantitative imaging of chlorophyll fluorescence. Planta 199, 573–582.
43. Scholes JD, Lee PJ, Horton P, Lewis DH. 1994. Invertaseunderstanding changes in the photosynthetic and

carbohydrate metabolism of barley leaves infected with powdery mildew. New Phytologist 126, 213–222.

44. Schreiber U. 2004. Pulse-Amplitude-Modulation (PAM) fluorometry and saturation pulse method. In: Papageorgiu G, Govindjee, eds. Chlorophyll a fluorescence: a signature of photosynthesis. Dordrecht, The Netherlands: Springer, 1–42an overview.
45. Schreiber U, Schliwa U, Bilger W. 1986. Continuous recording of photochemical and nonphotochemical chlorophyll fluorescence quenching with a new type of modulation fluorometer. Photosynthesis Research 10, 51–62.
46. Schwachtje J, Minchin PE, Jahnke S, van Dongen JT, Schittko U, Baldwin IT. 2006. SNF1-related kinases allow plants to tolerate herbivory by allocating carbon to roots. Proceedings of the National Academy of Sciences, USA 103, 12935–12940.
47. Siemens J, Keller I, Sarx J, Kunz S, Schuller A, Nagel W, Schmulling T, Parniske M, Ludwig-Muller J. 2006. Transcriptome analysis of Arabidopsis clubroots indicate a key role for cytokinins in disease development. Molecular Plant–Microbe Interactions 19, 480–494.
48. Sinha AK, Hofmann MG, Romer U, Kockenberger W, Elling L, Roitsch T. 2002. Metabolizable and non-metabolizable sugars activate different signal transduction pathways in tomato. Plant Physiology 128, 1480–1489.
49. Swarbrick PJ, Schulze-Lefert P, Scholes JD. 2006. Metabolic consequences of susceptibility and resistance in barley leaves challenged with powdery mildew. Plant, Cell and Environment 29, 1061–1076.
50. Tao Y, Xie Z, Chen W, Glazebrook J, Chang HS, Han B, Zhu T, Zou G, Katagiri F. 2003. Quantitative nature of Arabidopsis responses during compatible and incompatible interactions with the bacterial pathogen Pseudomonas syringae. The Plant Cell 15, 317–330.
51. Thomma BPHJ, Penninckx IA, Broekaert WF, Cammue BP. 2001. The complexity of disease signaling in Arabidopsis. Current Opinion in Immunology 13, 63–68.
52. van Kooten O, Snel JFH. 1990. The use of chlorophyll fluorescence

nomenclature in plant stress physiology. Photosynthesis Research 25, 147–150.

53. von Schaewen A, Stitt M, Schmidt R, Sonnewald U, Willmitzer L. 1990. Expression of a yeast-derived invertase in the cell wall of tobacco and Arabidopsis plants leads to accumulation of carbohydrate and inhibition of photosynthesis and strongly influences growth and phenotype of transgenic tobacco plants. EMBO Journal 9, 3033–3044.
54. Voegele RT, Wirsel S, Moll U, Lechner M, Mendgen K. 2006. Cloning and characterization of a novel invertase from the obligate biotroph Uromyces fabae and analysis of expression patterns of host and pathogen invertases in the course of infection. Molecular Plant–Microbe Interaction 19, 625–634.
55. Walters DR, McRoberts N. 2006. Plants and biotrophs: a pivotal role for cytokinins? Trends in Plant Science 11, 581–586.
56. Wright D, Baldwin B, Shephard M, Scholes J. 1995. Source–sink relationships in wheat leaves infected with powdery mildew. Physiological and Molecular Plant Pathology 47, 237–253.

Chapter 4

Domestication and Crop Physiology: Roots of Green-Revolution Wheat

J. Giles Waines* and Bahman Ehdaie

Department of Botany and Plant Sciences, University of California, Riverside, CA 92521-0124, USA

ABSTRACT

Background and Aims Most plant scientists, in contrast to animal scientists, study only half the organism, namely above-ground stems, leaves, flowers and fruits, and neglect below-ground roots. Yet all acknowledge roots are important for anchorage, water and nutrient uptake, and presumably components of yield. This paper investigates the relationship between domestication, and the root systems of landraces, and the parents of early, mid- and late green-revolution bread wheat cultivars. It compares the root system of bread wheat and 'Veery'-type wheat containing the 1RS translocation from rye.

Methods Wheat germplasm was grown in large pots in sand culture in replicated experiments. This allowed roots to be washed free to study root characters.

Key Results The three bread wheat parents of early green-revolution wheats have root biomass less than two-thirds the mean of some landrace wheats. Crossing early green-revolution wheat to an F_2 of 'Norin 10' and 'Brevor', further reduced root biomass in mid-generation semi-dwarf and dwarf wheats. Later-generation semi-dwarf wheats show genetic variation for root biomass, but some exhibit further reduction in root size. This is so for some California and UK wheats. The wheat–rye translocation in 'Kavkaz' for the short arm of chromosome 1 (1RS) increased root biomass and branching in cultivars that contained it.

Conclusions Root size of modern cultivars is small compared with that of landraces. Their root system may be too small for optimum uptake of water and nutrients and maximum grain yield. Optimum root size for grain yield has not been investigated in wheat or most crop plants. Use of 1RS and similar alien translocations may increase root biomass and grain yield significantly in irrigated and rain-fed conditions. Root characters may be integrated into components of yield analysis in wheat. Plant breeders may need to select directly for root characters.

INTRODUCTION

Domestication of plant species by humans in the last 10 000 years proceeded with observation and direct selection of mostly above-ground organs, namely stems, leaves, flowers and fruits. Below-ground organs, such as roots, were little observed, unless the root comprised a food-storage organ that was selected for directly. This general lack of interest in plant root systems, where up to half the plant was neglected, contrasts strongly with animals where the whole organism was observed and selected during domestication. Most plant scientists acknowledge roots are important for anchorage, and water and nutrient uptake from the soil solution. A small number consider the root system important for components of yield analysis (Bazzaz *et al.*, 2000). Bread wheat (*Triticum aestivum* L.) is an allohexaploid ($2n = 6x = 42$) with the genome formula BBAADD. It was formed from hybridization of domesticated tetraploid wheat [*T. turgidum* L. ssp. *dicoccum* (Schrank) Thell. genome formula BBAA] with diploid weedy goat grass (*Aegilops tauschii* Coss. DD) in north-west Iran or Armenia (Dvorak *et al.*, 1998). *Triticum turgidum* ssp. *dicoccum* is a domesticated race of wild emmer *T. turgidum* L. ssp. *dicoccoides* (Korn. ex Asch. & Graebn.) Thell. native to the fertile crescent of the Near East, and itself an allotetraploid formed from an ancestor of *Aegilops speltoides* Tausch as female parent (BB) and *T. urartu* Tum. ex Gandil. as male parent (AA) (Waines and Barnhart, 1992).

The history of the green revolution in wheat in the mid-20th century was outlined by Dr Norman Borlaug (Borlaug, 1968). This internationally important plant breeding programme was conducted largely by selection of above-ground organs. Wheat roots were rarely studied by breeders in the CIMMYT or national programmes. This paper investigates the relationship between domestication and the root system of bread wheat. It looks at root biomass of landraces and of the parent lines hybridized to select early-, mid- and late-generation green-revolution cultivars of bread wheat in Mexico, California and the UK.

This paper also compares root biomass of pure bread wheat and of 'Veery'-type wheat that contains the 1RS translocation from rye (*Secale cereale* L.). 'Veery' wheats are popular under drought or heat-stressed conditions for, on average, they may produce a 7 % grain yield advantage (Rajaram *et al.*, 1983). They are favoured where grain yield is more important than bread-making quality. Up to 50 % of all wheat cultivars in China may now carry the 'Veery' 1RS translocation (Zhou *et al.*, 2003). 'Veery' wheats may have the 1RS arm from two sources: the more common $1RS_{(K)}$ is from 'Kavkaz' winter wheat, and the original translocation was selected in Germany from crosses of wheat with a 'Petkus' rye for resistance to fungal diseases (Schlegel and Korzun, 1997), while the more recent $1RS_{(A)}$ arm is from 'Amigo' winter wheat and the translocation was selected in Oklahoma from crosses with Brazilian 'Insave' rye for aphid resistance (Sebesta and Wood, 1978). One of the first scientists to excavate and illustrate the root system of bread wheat and rye was Weaver (1926) who noted differences in length of seminal roots and lateral roots between the two species, but did not investigate cultivar differences within a species. Root growth differences among seven cultivars of Canadian spring bread wheat were reported byHurd (1968) who hybridized parents with large, deep root systems and other desirable characters to select new cultivars with larger root systems and higher grain yield. Mac Key (1973) was the first to observe roots and shoots of F_1 hybrid plants of spring tall cultivar 'Prins' and tall winter cultivar 'Starke' with 'Norin 10' and 'Tom Thumb' the source of the *Rht1*,*Rht2* and *Rht3* alleles used by Borlaug (1968) in his Mexican semi-dwarf wheat breeding programme. Mac Key (1973) noted 'Prins' and 'Starke' had larger root dry weights than the F_1 hybrids with 'Norin 10' or 'Tom Thumb'. He concluded that 'a tall wheat plant tends to have a deep, and a short wheat plant a shallow, root system'. This raised the possibility that breeding for semi-dwarf stems, controlled by the *Rht1* and *Rht2* alleles on homoeologous group 4 chromosomes, might condition semi-dwarf root systems that might affect adversely the amount of water and nutrients absorbed by the plant, and hence grain yield. This was investigated in isogenic lines of UK winter wheat by Lupton *et al.* (1974) and found not to be so. Ehdaie and Waines (1994) in California, and Mirrales *et al.* (1997) in Argentina, used isogenic lines of the Brazilian spring cultivar 'Maringa'. The semi-dwarf *Rht1* and *Rht 2* isolines had larger root biomass than the tall *rht* line. The dwarf *Rht3* isoline with the shortest stems also had the smallest root system. In the semi-dwarf *Rht1* and *Rht2* isolines, assimilates not used to develop a large shoot system might be diverted to the roots to develop a larger root system. Secondly, Mac Key (1973, table 5) observed large differences in root biomass among old and modern wheats. Old varieties such as German 'Brown Schlanstedt' had three times larger root biomass than dwarf Mexican 'Mayo 64' or dwarf Japanese 'Kohnosu 25' while Brazilian 'Frontana' and Mexican 'Pictic 62' were intermediate. Troughton and Whittington (1968) using cultivar chromosome substitution lines, observed that differences in root size between tall landrace 'Chinese Spring' with large root biomass, and cultivar 'Hope', with small root biomass, were largely governed by genetic factor(s) on chromosome 1A, which is a different genetic system from that controlling the *Rht* stem dwarfing genes on chromosome 4, which have a smaller effect on root size. (See Appendix for a full list of cultivars mentioned in the text, together with brief descriptions.)

Similarly, three old tall landraces from China and Iran had 2- to 4-fold larger root biomass than four semi-dwarf Mexican, Iraqi, Pakistani and California cultivars, all of which were descended from CIMMYT breeding material (Ehdaie *et al.*, 1991; Ehdaie and Waines, 1993, 1997; Ehdaie, 1995). The root biomass differences between the landraces and the CIMMYT-derived modern cultivars were always significantly different. The root size of other CIMMYT-derived bread wheats have since been determined and they are always significantly smaller than drought-tolerant landraces (J. G. Waines and B. Ehdaie, unpubl. res.), though admittedly only a few accessions have been observed. This tends to support the view that direct selection for only above-ground organs might also indirectly select for a small root system, especially under well-irrigated and well-fertilized growing conditions, as at CIMMYT breeding stations, where there would be no selection advantage for a larger root system.

What is lacking in this review of the few published observations on bread wheat root systems is knowledge of what happened, in a historic sense, in the progression from tall landraces to semi-dwarf and dwarf cultivars in the last 100 years. In barley (*Hordeum vulgare* L.) seedlings, Grando and Cecarelli (1995) noted an initial increase in root size from ten wild barleys to ten landrace barleys, then a marked decrease in root size from landraces to ten modern barley cultivars. Wild collections of bread wheat do not exist, only landraces. As far as we are aware, there is no report in the literature of the root systems of the three spring wheat parents whichBorlaug (1968) hybridized to select the first-generation, tall Mexican wheats such as 'Nainari 60' or 'Lerma Rojo', or the root system of the Japanese two-gene-dwarfed wheat 'Norin 10' that contributed the *Rht1*and *Rht2* stem-dwarfing genes to mid-generation semi-dwarf and dwarf cultivars released by CIMMYT. This research was undertaken to fill this void.

MATERIALS AND METHODS

Kernels of the three spring wheat parental cultivars, namely 'Gabo' from Australia, and 'Marroqui' and 'Mentana' from the Mediterranean region, hybridized by Borlaug (1968) to form early green-revolution bread wheats, and of Japanese, tall 'Aka Komugi' were obtained from Dr Harold Bockelman, USDA Small Grains Collection, Aberdeen, ID,

USA. Kernels of early Mexican tall 'Nainari 60' were obtained from Dr M. D. Gale, John Innes Centre, Norwich, UK, and mid-generation Mexican semi-dwarf 'Anza', 'Pavon 76' and 'Yecora Rojo' cultivars were obtained from Dr C. O. Qualset, Department of Plant Sciences, University of California, Davis. Kernels of late-generation Mexican wheats 'Bacanora', 'Pastor' and 'Rayon' were obtained from CIMMYT. Grains of the Pavon 1RS near isogenic lines, namely 'Pavon 76', 'Pavon $1RS_{(K)}$.1AL', 'Pavon $1RS_{(K)}$.1BL' and 'Pavon $1RS_{(K)}$.1DL' (Lukaszewski, 1993, 2000) and other Pavon translocation stocks were donated by Dr A. J. Lukaszewski, University of California, Riverside. Wheat kernels of similar weight were surface sterilized in a 10 % solution of commercial sodium hypochlorite, washed in distilled water, and allowed to germinate in Petri dishes. Seedling plants of similar primary leaf length and seedling root number were transplanted to black plastic pots containing a polyethylene bag with 5 kg washed sand #30 (Ehdaie *et al.*, 2003). A small hole in the bag allowed excess water to drain without sand being lost.

In an early experiment (Table 1; Ehdaie, 1995) plants were watered every 2–3 d and returned to field capacity in a closed sand culture system. CIMMYT semi-dwarf and dwarf wheats had root biomass values between 2·0 g and 2·5 g per plant in the sand-culture experiments, and 'Anza' with 2·35 g per plant was intermediate. Root biomass was expressed as a percentage of mid-generation CIMMYT wheat 'Anza'.

Table 1. Root biomass of landrace bread wheats from Iran and China, an old Californian cultivar and mid-green revolution cultivars from California, Iraq and Pakistan (data from Ehdaie, 1995, table 2)

Genotype	Dwarfing gene	Group	Root biomass (g)	% Anza	Height (cm)
Iran 49	*rht*	Landrace	6·43^{a*}	273	115
Chinese Spring	*rht*	Landrace	4·38^{b}	186	112
Iran 14	*rht*	Landrace	3·35^{c}	138	111
Ramona 50	?	CA cultivar	2·70^{cd}	114	108
Anza	*Rht1*	mid-gr†	2·35^{d}	100	88
Chenab 70	*Rht1*	mid-gr	2·43^{cd}	103	92

Sholeh	*Rht1*	mid-gr	2·30^{d}	97	100
Yecora Rojo	*Rht1/Rht2*	mid-gr	2·00^{d}	85	76

* Means followed by the same letters within a column are not significantly different at $P \leq 0{\cdot}05$, using an LSD test.

† mid-gr = mid-green revolution cultivar.

* Means followed by the same letters within a column are not significantly different at $P \leq 0{\cdot}05$, using an LSD test.

† mid-gr = mid-green revolution cultivar.

A second experiment was started in mid-January 2006, with 'Pavon 76' as the standard cultivar. UK winter cultivar 'Zion 19' requiring vernalization was given cold temperatures of 5 °C and 8 h light, for 6 weeks beginning 1 December 2005. Two ancestors of CIMMYT wheats, namely 'Norin 10' and 'Brevor' were initially included in the experiment, but their data were later removed because plants were extremely long-day photoperiodic and did not head, in the case of 'Brevor' at Riverside, until the end of July 2005, and even then poorly. Both wheats, which contributed the *Rht1* and *Rht2*stem dwarfing genes, were bred to be grown in northern Japan or in Washington State where summer day lengths may exceed 16 h. The genotypes grown in pots and for which data are reported are listed in Table 2. At physiological maturity, irrigation ceased and plants were allowed to dry. Shoots were separated from roots at the soil surface. Roots were gently washed free of sand and allowed to dry in a forced air oven at 60 °C for 7 d, then weighed.

Table 2. Root biomass of a Japanese bread wheat landrace, parents of early, mid- and late green revolution cultivars and two modern cultivars from California and Europe

Genotype	**Dwarfing gene**	**Group**	**Root biomass (g)**	**% Pavon 76**	**Height (cm)**
Aka Komugi	*rht*	Landrace	9·33^{a*}	186	130
Marroqui	*rht?*	Tunisia	6·38^{bcd}	127	91
Mentana	*rht?*	Italy	6·19^{bcd}	123	122
Gabo	*rht?*	Australia	4·00^{cd}	80	90
Nainari 60	*rht*	early-gr†	7·15^{ab}	142	100
Pavon 76	*Rht1*	mid-gr	5·00^{bcd}	100	76

Bacanora	*Rht1*	late-gr	6·22^{bc}	124	87
Rayon	*Rht1*	late-gr	4·17^{cd}	83	76
Pastor	*Rht1*	late-gr	3·69^{d}	74	70
Summit	*Rht1*	late CA	4·49^{cd}	90	68
Zion 19	*Rht8*	late UK	4·61^{cd}	92	72

* Means followed by the same letters within a column are not significantly different at $P \leq 0{\cdot}05$, using an LSD test.

† early-gr = early-green revolution cultivar; mid-gr = mid-green revolution cultivar; late-gr = late-green revolution cultivar.

* Means followed by the same letters within a column are not significantly different at $P \leq 0{\cdot}05$, using an LSD test.

† early-gr = early-green revolution cultivar; mid-gr = mid-green revolution cultivar; late-gr = late-green revolution cultivar.

In the second experiment (Table 2) plants were watered generally each day as needed and allowed to drain. All root biomass values were expressed as a percentage of 'Pavon 76', also a mid-green-revolution wheat similar to 'Anza'. Further, 'Pavon 76' was chosen as the standard CIMMYT cultivar by Dr A. J. Lukaszewski and has replaced landrace 'Chinese Spring' as the cytogenetic-stock wheat. Wheat breeders do not like the morphological or quality characters of 'Chinese Spring', whereas they do like 'Pavon 76', which is a semi-dwarf CIMMYT wheat, of high grain quality that is also cytogenetically tractable. The rye 1RS translocations are in the background of 'Pavon 76' wheat.

Root biomass and grain yield of the 'Pavon 76' near-isogenic line series that compared bread wheat with 1RS$_{(K)}$ 'Veery' wheats, in glasshouse and field experiments, were reported by Ehdaie *et al.* (2003). Experiments that investigated the effect of the 1RS$_{(K)}$ arm in tetraploid wheat and compared root biomass of 'Aconchi' durum and 'Aconchi 1RS$_{(K)}$.1BL', and that investigated the effect of different sources of the 1RS arm on root biomass in 'Pavon 1RS$_{(A)}$.1AL' and 'Pavon 1RS$_{(K)}$.1AL' were reported earlier (Waineset *al.*, 2004; Waines and Ehdaie, 2005).

RESULTS

Mexican wheats

In the first experiment conducted in 1993, the three tall bread wheat landraces, presumed to carry a *rht* allele, though not necessarily the same one, had root biomass ranging from 3·25 g (Iran 14) to 6·43 g (Iran 49) with a mean root biomass of 4·68 g (Table 1). Plant height for the same genotypes ranged from 111 cm to 115 cm. 'Ramona 50', a popular California wheat released before the introduction of Mexican wheats, had root biomass smaller than the landraces, but larger than four mid-generation, green-revolution wheats released in California, Iraq and Pakistan. The cultivar with the smallest root biomass was the two-gene (*Rht1/Rht2*) dwarf 'Yecora Rojo', which also had the shortest stem height at 76 cm.

In the second experiment conducted in 2006 (Table 2) root biomass values were larger than those reported in Table 1, presumably because the plants received twice as much water (25 kg *vs*. 12·74 kg). However, the same trend was observed. The tall (137 cm) Japanese landrace 'Aka Komugi' also had the largest root biomass (9·33 g). The parents of early-generation Mexican wheats had root biomass that ranged from 4·01 g ('Gabo') to 6·38 g ('Marroqui') with a mean of 5·52 g. Plant height of these lines was similar (119–122 cm). The tall (100 cm) (*rht*) early-generation Mexican wheat 'Nainari 60', with root biomass of 7·15 g was selected from segregating hybrid populations of the three parents, which root biomass, but not plant height, it exceeded. Early-generation Mexican wheats were crossed to segregating F_2 lines of 'Norin 10' ×'Brevor', whose root biomass it was not possible to measure because both cultivars were day-length sensitive for flowering at Riverside. The mid-generation green-revolution wheat, 'Pavon 76', that carried the semi-dwarf *Rht1* allele had root biomass of 5·01 g and plant height of 87 cm. Late-generation semi-dwarf green-revolution wheats that carried the *Rht1* allele had variable root biomass that ranged from 'Bacanora' with 6·22 g and 86 cm, to 'Pastor' with 3·69 g and 70 cm and a mean root biomass of 4·69 g. A recent California semi-dwarf (*Rht1*) wheat 'Summit' had root biomass of 4·49 g and height of 68 cm, and a recent semi-dwarf UK winter wheat 'Zion 19' had root biomass of 4·60 g and height of 72 cm.

'Veery' wheats

Earlier, experiments were undertaken to elucidate the morphological and/or physiological basis of 'wide adaptation' in 'Veery'-type wheats carrying the 1RS arm translocated for wheat chromosome arm 1BS or 1AS. That this might involve differences in root biomass was supported by use of near-isogenic lines where plants with the 1RS translocation always had larger root biomass and branching in pot cultures (Fig. 1). In this experiment, where pots were weighed every 2–3 d and returned to field capacity, CIMMYT bread wheat 'Pavon 76' had root biomass of 2·5 g, while near-isogenic lines for the $1RS_K$ arm had in 'Pavon 1RS.1AL' 3·4 g, 'Pavon 1RS.1BL' 3·0 g and 'Pavon 1RS.1DL' 3·4 g root biomass, respectively, in well-watered pot cultures. Root biomass was similar to results obtained in Table 1. In drought stressed pot cultures root biomass showed a similar trend. In field experiments, grain yield under irrigation was positively associated with root biomass in glasshouse sand cultures (Ehdaie *et al.*, 2003).

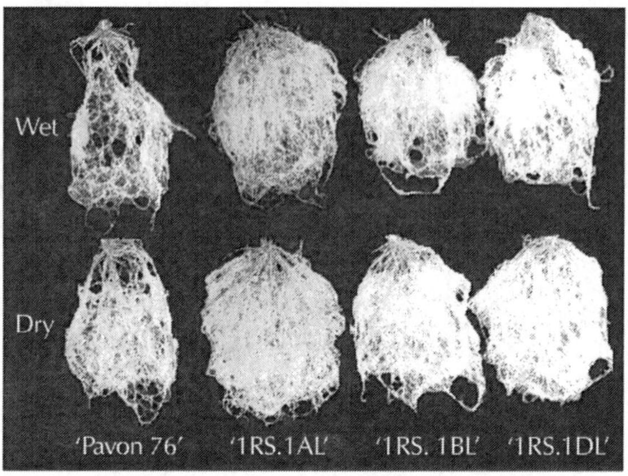

Figure. 1. Root biomass of green-revolution bread wheat 'Pavon 76' and its near isogenic rye 1RS translocation lines 'Pavon 1RS.1AL', 'Pavon 1RS.1BL' and 'Pavon 1RS.1DL' grown under well-watered (wet) and drought-stressed (dry) pot sand-culture conditions in a cooled glasshouse.

In another experiment, the root biomass of 'Aconchi' durum was 3·17 g per plant whereas that of 'Aconchi $1RS_{(K)}$.1BL' was 4·41 g per plant and showed a 39 % increase (Waines *et al.*, 2004). In a separate experiment, root biomass of 'Pavon $1RS_{(A)}$.$1AL_{(A)}$' showed a 9

% increase over that of 'Pavon 76', while 'Pavon $1RS_{(K)}.1AL_{(P)}$' showed a 31 % increase over the 'Pavon 76' control (Waines *et al.*, 2004; Waines and Ehdaie, 2005).

DISCUSSION

In these pot, sand culture studies, tall landrace bread wheats had larger root biomass than mid-green revolution wheats (Tables 1 and 2). These results were confirmed independently using field soil and sand cultures in Indiana by Crowley *et al.* (2005, 2006) who found 'Iran 49' always had more root biomass than 'Iran 14', and some local bread wheat cultivars. In contrast, the three parents of Mexican wheat listed by Borlaug (1968) have a mean root biomass of 5·52 g well below landrace 'Aka Komugi'. 'Gabo' from Australia had an especially small root system, relative to the other parents 'Mentana' and 'Marroqui'. CIMMYT's early-generation 'Nainari 60' had root biomass (7·15 g) that exceeded its parents.

Crossing the early-generation Mexican wheats with F_2 segregants of 'Norin 10' and 'Brevor' to introduce the stem-dwarfing *Rht1* and *Rht2*alleles did not increase the root biomass of the mid-generation green-revolution wheat such as 'Pavon 76' (Table 2), which has root biomass below the mean of the parents (5·52 g). The results are, by implication, similar in Table 1. Further hybridization and selection of late green-revolution wheats from the CIMMYT programme did introduce some variability in the case of 'Bacanora', but 'Rayon' and 'Pastor' show a continued decrease in root biomass. A similar trend of unconsciously selecting wheats with small root biomass, as a consequence of selection under optimum growing conditions, is evident in two randomly chosen cultivars, modern Californian 'Summit' and United Kingdom 'Zion 19' bread wheat. As 'Zion 19' is a winter wheat, there may have been similar unconscious selection of small root biomass in winter wheat as well as spring wheat.

'Ramona 50' bread wheat was released in California in 1950 before the introduction of CIMMYT germplasm. It has a root biomass of 2·70 g per plant. Thus the selection of one cultivar with a small root system occurred in the USA before the introduction of germplasm from the CIMMYT programme in Mexico. Gabo bread wheat from Australia is another example of a cultivar released around 1945 that has a small root system and is adapted to Australia's summer rainfall regime. One of the first reports of genetic control of cultivar differences for root biomass in bread wheat was by Troughton and Whittington (1968), andMonyo and Whittington (1970) who observed landrace 'Chinese Spring' had root biomass twice that of 1930 cultivar 'Hope'. The difference, discerned by use of cultivar whole-chromosome substitution lines, was controlled by gene(s) on chromosome 1A. The controlling factor(s) were not mapped to chromosome arms, nor does this information appear to have been much used in wheat breeding programmes with the possible exception of that of Hurd (1971) in Canada.

No evidence was found that Borlaug (1968) and the CIMMYT wheat breeding team considered root size or other root characters in the parents or offspring of their breeding programmes in the 1940s through 1970s. In this respect, plant breeding is different from animal breeding where healthy feet and strong legs were observed and selected for in breeding programs throughout animal domestication. Most wheat breeding programmes involve only above-ground plant organs that are readily observed and selected. With a few exceptions (Hurd *et al.*, 1972;Hurd, 1974; Richards and Passioura, 1989) root characters were not observed or selected for. The need to breed for root characters was advocated by Mac Key (1973, 1978) who reported a positive relationship between use of *Rht* stem dwarfing alleles and small root size. Even though later workers did not confirm this relationship (Lupton *et al.*, 1974;Mirrales *et al.*, 1997), investigation of a possible grain yield advantage by selecting for a larger root system has been slow. Part of the problem was that there are few near-isogenic line series in bread wheat with known differences in root size that might be used to demonstrate the advantage of a larger root size. Another reason may be that many plant scientists do not like to work with roots because root measurements are time consuming and it is difficult to measure root traits for a large number of breeding lines.

One near-isogenic set is material developed at CIMMYT which differs for presence and absence of the 'Veery $1RS_{(K)}$.1BL' translocation in the same genetic background. This translocation from winter wheat 'Kavkaz', involves the short arm of chromosome 1 of rye translocated for the short arm of chromosome 1B or 1A of bread wheat, and hence may involve the same genetic system identified by Troughton and Whittington (1968). Field research by Manske and Vlek (2002) at CIMMYT in Mexico was the first to report that wheat genotypes containing 1RS had thinner roots and higher root-length density compared with their 1BS checks. In glasshouse pot studies involving 1RS isolines of 'Pavon 76', the lines containing $1RS_{(K)}$.1AL, $1RS_{(K)}$.1BL and $1RS_{(K)}$.1DL always had more root biomass and root branching than 'Pavon 76' (Fig. 1; Ehdaie *et al.*, 2003). Root biomass and grain yield were positively correlated under well-watered and droughted pot conditions. In companion field studies, lines containing 1RS also yielded more grain in a well-irrigated treatment, but not in a drought-stressed treatment. The stress tolerance index, calculated from field grain yield and kernel weight, indicated 'Pavon 76' isolines for $1RS_{(K)}$ had greater tolerance to stressful environments than 'Pavon 76' (Ehdaie *et al.*, 2003). Based on these pot and field studies those authors concluded the greater adaptability of certain 1RS.1BL translocation lines reported in the literature may be due to their greater root biomass and higher transpiration rate. Root biomass was not investigated in the field experiments. However, grain yield measured in the field experiments was positively correlated with root biomass measured in the pot experiments (Ehdaie and Waines, 2003). If root biomass is increased in the field, these results might be in agreement with the model developed for winter wheat in the UK by King *et al.* (2003), who predicted more grain yield from a wheat plant with a deep root system that was able to mine a larger soil profile for more water and nutrients than a shallow root system mining a smaller soil profile.

Comparing CIMMYT semi-dwarf durum 'Aconchi 1BS.1BL' with the near isogenic durum 'Aconchi $1RS_{(K).}$1BL' indicated 1RS increased root biomass by 39 % in tetraploid wheat. Hence, the 1RS effect has similar root biomass results in tetraploid and hexaploid wheat. However, the 1RS arm, in 'Aconchi' durum background, carries a gene that also conditions partial floret sterility (Waines *et al.*, 2004).

Comparing different sources of the 1RS arm from 'Kavkaz' or from 'Amigo' in 'Pavon 76' bread wheat background, root biomass of 'Pavon $1RS_{(A).}1AL_{(A)}$' was 9 % more than 'Pavon 76', whereas 'Pavon $1RS_{(K)}.1AL_{(P)}$' was 31 % more than 'Pavon 76' (Waines *et al.*, 2004). Therefore in 'Pavon 76' background, the 'Kavkaz' 1RS arm has a 3·4 times larger effect on root biomass than the 'Amigo' 1RS arm, which suggests there may be genetic variation for root size within rye.

It was assumed that genotypes containing 1RS that increase root biomass in sand cultures will also increase root biomass in field soils, though not necessarily to the same extent. This is supported by the field results ofManske and Vlek (2002) and the sand pot and field results of Ehdaie *et al.*(2003). The general increase in grain yield reported in 'Veery' wheats containing 1RS may not be attributed solely to increase in root biomass. Research in the UK reports isogenic lines with the 1RS arm may also store more water-soluble carbohydrates in the stems before anthesis and mobilize more water-soluble carbohydrates to grains during grain filling (Shearman *et al.*, 2005). The increased production of water-soluble carbohydrates may be due to increased uptake of water and nutrients, as the model of King *et al.* (2003) suggests, and/or to increased photosynthesis, with carbohydrates produced before flowering stored in the stem before being re-mobilized after flowering to fill the grains.

This research demonstrated the parental genotypes used to breed early-generation green-revolution wheats have small root biomass in sand cultures, especially 'Gabo' from Australia. This may be controlled by genes on homoeologous chromosome group 1, rather than by the *Rht* dwarfing genes on homoeologous group 4. This interpretation is supported by theF_2 segregation data from a cross of tall 'Chinese Spring' with large root biomass and dwarf 'Yecora Rojo' with small root biomass where plants with small root biomass and tall stems or large roots and dwarf stems were obtained (Ehdaie and Waines, 1993).

Some semi-dwarf mid- and later-generation green-revolution cultivars selected in Mexico, California and the UK also have small root systems in sand cultures. The root systems of some modern bread wheat cultivars appear to be getting smaller or at least not increasing. The reduction in root size began before introduction of green-revolution wheats and may be a general result of domestication and breeding (Mac Key, 1978). In the case of Mexican wheats, this may also be the result of unconscious selection for increased grain yield in irrigated and well-fertilized conditions. Small root systems may account for why some green-revolution wheats perform well in optimum conditions, but poorly in drought and heat-stressed conditions. Alternatively, the parents presently used in crossing blocks may all have similar small root size, and with little available genetic variation, there is no ability to select for a larger root system. There may be reluctance on the part of crop physiologists and breeders to experiment with near-isogenic variation in root size and to investigate its effect on water and fertilizer uptake and grain yield. The lack of root studies contrasts with the many experiments on the effect of near-isogenic variation for shoot height on grain yield in bread wheat.

There has been little characterization of differences in the root systems of wild diploid and tetraploid wheats and their relatives, or of landraces of diploid, tetraploid or hexaploid wheats by germplasm curators. However, this germplasm is recognized as a source of potential drought-adaptive traits for wheat breeding programmes (Reynolds *et al.*, 2007). Also, the possibility of using plant-adaptive mechanism, including genetic variation in wheat root systems, to breed crops for drought and salinity-prone environments has been considered (Reynolds *et al.*, 2005).

In conclusion, presence of $1RS_K$, from 'Petkus' rye, and $1RS_A$ from 'Insave' rye, associates with larger root biomass and branching than 'Pavon 76' hexaploid wheat. Similarly, presence of the $1RS_K.1BL$ translocation in 'Aconchi' increased root biomass over that of 'Aconchi' tetraploid wheat. The 1RS arm may increase root biomass in other tetraploid and hexaploid wheats. Grain yield was associated with increased root biomass in the 'Pavon 76' near-isogenic lines with or without the 1RS arm. This may open a way to improve root characters, grain yield and ultimately grain quality in bread wheat. If a larger root biomass increases water and nitrogen absorption, this may also suggest a way to reduce nitrate pollution. These results raise the question is the root size of modern wheat large enough for maximum yield? Would more root biomass and branching increase grain yield of pure wheats with semi-dwarf stems in irrigated and rain-fed conditions? Has unconscious selection for small root characters been taken too far in wheat domestication and breeding? The shallow soils reported in Australia, with high boron in the subsoil and largely spring–summer rainfall, suggests not all environments might benefit from a larger root system. However, wheat grown in deeper soils and in other areas such as California and Western Europe, may benefit from a larger root system. To optimize grain yield, it may be necessary to tailor the root system to the soil. At present, root characters are not considered important for components of grain yield analysis in wheat (Donald, 1968), even though a general model for all plants that includes both shoot and root biomass has been published (Bazzaz *et al.*, 2000). Crop physiologists and plant breeders may need to design the complete wheat plant, including roots and shoots, not only the organs observed above ground.

ACKNOWLEDGEMENTS

Funding to pay the Open Access publication charges for this article was provided by the OECD.

APPENDIX

List of species and/or of bread wheat cultivars mentioned in the text, their origin and pertinent comments.

'**Aconchi**': semi-dwarf (*Rht1*), spring, CIMMYT, durum released in Mexico.

'**Aconchi 1RS$_K$.1BL**': durum 'Veery' near-isogenic line.

'**Akakomugi Tall**': very tall (*rht*) plants in seed of dwarf Japanese 'Akakomugi' received from USDA Small-Grains Germplasm Collection, Aberdeen, ID.

'**Amigo**': Oklahoma winter wheat contains 1RS.1AL 'Veery' translocation from 'Insave' rye.

'**Anza**': semi-dwarf (*Rht1*) CIMMYT-derived bread wheat released in California and Australia.

'**Bacanora**': semi-dwarf (*Rht1*), spring, late-generation CIMMYT bread wheat, Mexico.

'**Brevor 51**': Vogel's semi-dwarf, winter long-day responsive cultivar with stiff straw that allowed increased nitrogen rates and higher grain yields in Pacific Northwest, USA.

'**Brown Schlanstedt**': selected from landrace, end of 19th century, Germany.

'**Chenab 70**': semi-dwarf (*Rht1*) CIMMYT-derived bread wheat released in Pakistan.

'**Chinese Spring**': 1940 selected from Chinese wheat landrace by E. R. Sears.

'**Frontana**': 1945, Brazilian bread wheat cultivar tolerant of soil acidity.

'**Gabo**': 1945, tall bread wheat, Australia.

'**Hope**': 1935, winter wheat cultivar, USA, with small root system relative to 'Chinese Spring'.

'**Insave**': rye (*Secale cereale*) cultivar released in Brazil.

'**Iran 14**': tall (*rht*) selection out of Iranian Khouzistan bread wheat landrace by Ehdaie.

'**Iran 49**': tall (*rht*) selection out of Iranian Baluchistan bread wheat landrace by Ehdaie.

'**Kavkaz**': Siberian winter bread wheat contains 1RS.1BL 'Veery' translocation from 'Petkus' rye.

'**Kohnosu 25**': bread wheat cultivar released by the Kohnosu station near Tokyo, Japan.

'**Lerma Rojo 64**': early tall (*rht*) Mexican spring bread wheat cultivar released by CIMMYT.

'**Maringa**' (IAC-5): 1970, tall Brazilian bread wheat cultivar

'**Marroqui**' ('Florence Aurore'): Tunisia, spring bread wheat grown in North Africa.

'**Mayo 64**': CIMMYT semi-dwarf bread wheat.

'**Mentana**': 1930, tall, disease-resistant, Italian bread wheat.

'**Nainari 60**': early tall (*rht*) Mexican spring bread wheat cultivar released by CIMMYT.

'**Norin 10**': dwarf Japanese cultivar, used as source of *Rht1* and *Rht2* genes in USA after 1946 and Mexican wheat breeding programmes after 1950.

'**Pavon 76**': semi-dwarf (*Rht1*) spring CIMMYT bread wheat released in Mexico.

'**Pavon $1RS_K.1AL$**': Veery (Kavkaz) near-isogenic line of 'Pavon 76'.

'**Pavon $1RS_K.1BL$**': Veery (Kavkaz) near-isogenic line of 'Pavon 76'.

'**Pavon $1RS_K.1DL$**': Veery (Kavkaz) near-isogenic line of 'Pavon 76'.

'**Pavon $1RS_A.1AL$**': Veery (Amigo) near-isogenic line of 'Pavon 76'.

'**Pastor**': semi-dwarf (*Rht1*) spring, late-generation CIMMYT bread wheat, Mexico.

'**Petkus**': rye (*Secale cereale*) one of several cultivars released commercially in Germany.

'**Pitic 62**': CIMMYT semi-dwarf bread wheat cultivar.

'**Prins**': tall Swedish spring bread wheat.

'**Ramona 50**': tall (*rht*) spring bread wheat from California.

'**Rayon**': semi-dwarf (*Rht1*) spring, late-generation CIMMYT bread wheat, Mexico.

'**Sholeh**': semi-dwarf, CIMMYT-derived bread wheat introduced to Khouzistan from Iraq.

'**Starke**': tall Swedish winter bread wheat.

'**Summit**': modern commercial semi-dwarf spring bread wheat released in California.

'**Tom Thumb**': very dwarf cultivar, source of *Rht3* gene. Not used commercially.

'**Yecora Rojo 70**': two-gene dwarf (*Rht1, Rht2*) CYMMYT spring bread wheat released in Mexico, still grown in southern California/Arizona, with small root biomass.

'**Zion 19**': modern UK commercial semi-dwarf (*Rht8*) winter bread wheat.

REFERENCES

1. Bazzaz FA, Ackerly DD, Reekie EG. Reproductive allocation in plants. In: Fenner M, editor. Seeds: the ecology and regeneration of plant communities. 2nd edn. Wallingford: CAB International; 2000. p. 1-29.
2. Borlaug NE. Wheat breeding and its impact on world food supply. In: Finlay KW, Shepherd KW, editors. Proceedings of the 3rd International Wheat Genetics Symposium. Canberra. Sydney: Australian Academy of Sciences/Butterworths; 1968. p. 1-36.
3. Crowley NA, Uphause J, Ohm H. Agronomy Abstracts 8–5. Salt Lake City, UT, USA; 2005. Wheat lines with different root volume. 6 November.
4. Crowley NA, Uphause J, Ohm H. Agronomy Abstracts 10–4. Indianapolis, IN, USA; 2006. Identification of wheat genotypes

with large root volume to potentially increase drought tolerance. 12 November.

5. Donald CM. The design of wheat ideotypes. In: Finlay KW, Shepherd KW, editors. Proceedings of the 3rd International Wheat Genetics Symposium. Sydney, Australia: Butterworths; 1968. p. 377-387.
6. Dvorak J, Luo MC, Yang Z-L. Genetic evidence on the origin of Triticum aestivum. In: Damania AB, Valkoun J, Willcox G, Qualset CO, editors. The origins of agriculture and crop domestication. Aleppo, Syria: ICARDA; 1998. p. 235-251.
7. Grando S, Cecarelli S. Seminal root morphology and coleoptile length in wild (Hordeum vulgare ssp. spontaneum) and cultivated (H. vulgare ssp. vulgare) barley. Euphytica 1995;86:73-80.
8. Ehdaie B. Variation in water-use efficiency and its components in wheat. II. Pot and field experiments. Crop Science 1995;35:1617-1626.
9. Ehdaie B, Waines JG. Variation in water-use efficiency and its components in wheat. I. Well-watered pot experiment. Crop Science 1993;33:294-299.
10. Ehdaie B, Waines JG. Growth and transpiration efficiency of near isogenic lines for height in a spring wheat. Crop Science 1994;34:1443-1451.
11. Ehdaie B, Waines JG. Growth and evapotranspiration efficiency in landrace and dwarf spring wheats. Journal of Genetics and Breeding 1997;51:201-209.
12. Ehdaie B, Waines JG. 1RS translocation increases root biomass in Veery-type wheat isogenic lines and associates with grain yield. In: Pogna NE, Romano M, Pogna EA, Galterio G, editors. Proceedings of the 10th International Wheat Genetics Symposium. Paestum, Italy: Rome. ISC; 2003. p. 693-695.
13. Ehdaie B, Hall AE, Farquhar GD, Nguyen HT, Waines JG. Water-use efficiency and carbon isotope discrimination in wheat. Crop Science 1991;31:1282-1288.
14. Ehdaie B, Whitkus RW, Waines JG. Root biomass, water-use efficiency and performance of wheat–rye translocations of chromosome 1 and 2 in spring bread wheat Pavon. Crop Science 2003;43:710-717.

15. Hurd EA. Growth of roots of seven varieties of spring wheat at high and low moisture levels. Agronomy Journal 1968;60:201-205.
16. Hurd EA. Can we breed for drought resistance? In: Larson KL, Easton JD, editors. Drought injury and resistance in crops. 1971. p. 77-88. Crop Science Society Publication no. 2.
17. Hurd EA. Phenotype and drought tolerance in wheat. Agricultural Meteorology 1974;14:39-55.
18. Hurd EA, Townley-Smith TF, Patterson LA, Owen CH. Techniques used in producing Wascana wheat. Canadian Journal of Plant Science 1972;52:689-691.
19. King J, Gay A, Sylvester-Bradley R, Bingham I, Foulkes J, Gregory P, et al. Modelling cereal root systems for water and nitrogen capture: towards an economic optimum. Annals of Botany 2003;91:383-390.
20. Lukaszewski AJ. Reconstruction in wheat of complete chromosome 1B and 1R from 1RS.1BL translocation of Kavkaz origin. Genome 1993;36:821-824.
21. Lukaszewski AJ. Manipulation of the 1RS.1BL translocation in wheat by induced homoeologous recombination. Crop Science 2000;40:216-225.
22. Lupton FGH, Oliver RH, Ellis FB. Root and shoot growth in semidwarf and taller winter wheats. Annals of Applied Biology 1974;77:129-144.
23. Mac Key J. The wheat root. In: Sears ER, Sears LMS, editors. Proceedings 4th International Wheat Genetics Symposium. Columbia, Missouri, USA; 1973. p. 827-842.
24. Mac Key J. Wheat domestication as a shoot : root interrelation process. In: Ramanujam S, editor. Proceedings 5th International Wheat Genetics Symposium. New Delhi: Indian Agricultural Research Institute; 1978. p. 875-890.
25. Manske GGB, Vlek PLG. Root architecture-wheat as a model plant. In: Waisel Y, Eshel A, Kafkafi U, editors. Plant roots: the hidden half. 3rd edn. New York, NY: Marcel Dekker; 2002. p. 249-259.

26. Mirrales DJ, Slafer GA, Lynch V. Rooting patterns in near isogenic lines of spring wheat for dwarfism. Plant and Soil 1997;197:79-86.

27. Monyo JH, Whittington WJ. Genetic analysis of root growth in wheat. Journal of Agricultural Science, Cambridge 1970;74:329-338.

28. Rajaram S, Mann CE, Ortiz-Ferrara G, Mujeeb-Kazi A. Adaptation, stability and high yield potential of certain 1B/1R CIMMYT wheats. In: Sakamoto S, editor. Proceedings of the 6th International Wheat Genetics Symposium. Kyoto, Japan; 1983. p. 613-621.

29. Reynolds M, Dreccer F, Trethowan R. Drought adaptive traits derived from wheat wild relatives and landraces. Journal of Experimental Botany 2007;58:177-186.

30. Reynolds MP, Mujeeb-Kazi A, Sawkins M. Annals of Applied Biology. Vol. 146. 2005. Prospects for utilising plant-adaptive mechanisms to improve wheat and other crops in drought- and salinity-prone environments; p. 239-259.

31. Richards RA, Pasioura JB. A breeding program to reduce the diameter of the major xylem vessel in the seminal roots of wheat and its effect on grain yield in rain-fed environments. Australian Journal of Agricultural Research 1989;40:943-950.

32. Schlegel R, Korzun V. About the origin of 1RS.1BL wheat-rye chromosome translocations from Germany. Plant Breeding 1997;116:537-540.

33. Sebesta EE, Wood EA. Agronomy Abstracts. Madison, WI: ASA; 1978. Transfer of greenbug resistance from rye to wheat with X-rays; p. 61-62.

34. Shearman VJ, Sylvester-Bradley R, Scott RK Foulkes MJ. Physiological processes associated with wheat yield progress in the UK. Crop Science 2005;45:175-185.

35. Troughton A, Whittington WJ. The significance of genetic variation in root systems. In: Whittington WJ, editor. Root growth. Proceedings of the 15th Easter School in Agricultural Sciences. New York, NY: University of Nottingham; 1968. p. 296-314. Plenum.

36. Waines JG, Barnhart D. Biosystematic research in Aegilops and

Triticum. In: Seberg O, Lundquist A, editors. Proceedings of the 1st International Triticeae Symposium. Helsingborg, Sweden: Hereditas Offprint; 1992. p. 207-212.

37. Waines JG, Ehdaie B. Optimizing root characters and grain yield in wheat. In: Czech Journal of Genetics and Breeding; Proceedings of the 5th International Triticeae Symposium; Prague, Czech Republic. 2005.41(Special Issue) p. 326-330.
38. Waines JG, Ehdaie B, Sharma S. Crop Science Society of America Annual Meeting, Poster Abstract 1108. Seattle, WA; 2004. Effect of origin of 1RS translocation on root biomass in wheats.
39. Weaver JE. Root development of field crops. New York, NY: McGraw-Hill; 1926.
40. Zhou Y, He ZH, Liu JJ, Liu L. Distribution of 1BL/1RS translocation in Chinese winter wheat and its effect on noodle quality. In: Pogna NE, Romano M, Pogna E, Galterio G, editors. Proceedings 10th International Wheat Genetics Symposium. Paestum Italy; 2003. p. 1419-1421.

Chapter 5

Fruit Transpiration: Mechanisms and Significance for Fruit Nutrition and Growth

Giuseppe Montanaro, Bartolomeo Dichio and Cristos Xiloyannis

Department of Crop Systems, Forestry and Environmental Sciences University of Basilicata Italy

INTRODUCTION

Water and minerals transport in plant occur through the transpiration stream as triggered mainly by the environmental conditions. The transpiration of leaves has been studied in great detail over many years and the roles played by the various leaf structures (cuticle, stomata, lenticels etc) are well understood. Similarly well understood are the influences on foliar transpiration of meteorological variables such as temperature, radiation, vapour pressure deficit and wind, also, how these variables can be differently important to foliar transpiration depending on whether one is considering an isolated leaf or an entire canopy (Jarvis, 1985). In contrast with this, the transpiration of fruits has not been studied to nearly the same extent, neither with regard to skin structures and their associated functional properties nor with regard to the transpiration response of fruit to the various meteorological variables.

Most fruit-crop species (including apple, apricot, avocado, kiwifruit, capsicum and tomato), suffer from pre- and post-harvest physiological disorders. Higher incidences of these physiological disorders have many times been reported as being associated with lower concentrations of calcium (Ca) in the fruits (Faust et al. 1968, Tzoutzoukou and Bouranis 1997, Ferguson et al. 2003, Thorpe et al. 2003). Transpiration is the main driving force for the xylem stream (White and Broadley 2003) in which Ca seems to move relatively freely while this ion is also well known to be substantially immobile in the phloem (Buckowak and Wittwer 1957). Along with the observation that fruit are largely phloem fed, calcium's well know xylem mobility and phloem immobility explain in part why fruit are generally low-Ca organs and also why higher fruit transpiration rates are sometimes associated with increased fruit Ca levels (Cline and Hanson 1992, Tromp and Van Vuure 1993, Montanaro et al 2006 and 2010). It is therefore reasonable to hypothesise that for any particular fruit, the seasonal integral of fruit transpiration rate will predict fruit Ca content at harvest and thus (at least potentially) the incidence of Ca-related physiological disorders. This chapter focuses the mechanisms behind fruit transpiration in some fruit tree species, its seasonal trend and discusses the significance of fruit water loss on mineral composition particularly on Ca accumulation.

UNDERSTANDING AND MEASURING FRUIT TRANSPIRATION

Based on the Fick's- or Ohm's-law analogy (Fiscus et al., 1982), specific fruit transpiration rate (T_F) (the water vapour efflux per unit area of skin surface) depends on the product of the conductance properties (C) of the fruit skin and the driving force (V):

$$T_F = C \times V$$

Fruit conductance depends on skin properties determining its permeance, hence it is species specific. In addition it varies along with the season. Smith et al (1985) showed that the skin conductance in kiwifruit berry decreases rapidly during the early two weeks after anthesis (Fig. 1), thereafter the conductance decreases again reaching the minimum at about 8 weeks after anthesis where remains relatively constant until harvest.

Back to the Eq. 1, the term V is the difference in water vapour pressure between the airspaces inside the fruit and the air immediately outside it. The driving force depends on environmental conditions such as temperature, radiation, relative humidity, etc. Under most commercial training systems in most of fruits, their intercellular air (which is very close to saturation with relative humidity values close to ~99.4% (Nobel, 2005)) will have a water vapour pressure very close to the theoretical value for water at the temperature of this intercellular air. This means that transpiration will be driven substantially by a vapour pressure gradient whose value is numerically close to the water vapour pressure deficit (VPD) (Fig. 2) of the bulk atmosphere and this in turn is a reasonably exact function just of the surrounding air's temperature and of its relative humidity (Goudriaan and van Laar 1994). Based on Eq. 1, fruits under higher relative humidity are predicted to have a lower transpiration. Results presented by Li et al. (2002), even in a preliminary form, are in line with this prediction. In that study, it has been reported that an increase of relative humidity from 40 to 60% induced a reduction of fruit water loss by approximately 30-50% in two peach varieties under laboratory condition.

Measurement of fruit water loss could be performed in both attached and detached fruit. Regular leaf gas exchange instrumentation equipped with appropriate chamber helps to measure fruit transpiration in attached fruit, particularly for those species (e.g. kiwifruit) with relatively long peduncle (Photo 1). Since such a instrument requires knowledge of the surface area being analysed, it is required to preliminarily determine an appropriate correlation with fruit area and some geometrical traits of fruit which could be easily measured in the field (e.g. length, diameter). For example, for estimates of fruit surface area in kiwifruit berry a fruit sample (×70) were carefully peeled during the growing season and their skin area measured using a portable area meter (Model Li-3000, LI-COR). For the same fruit, estimated fruit surface area (Es) was calculated as ES = L*W*3.14, where L was the fruit length and W the maximum fruit diameter. The fitted linear regression ($y = 1.0078 * ES + 0.798$; $R^2 = 0.97$) between estimated and measured fruit surface area was then used in the field for measurements of fruit transpiration (Montanaro et al., 2006).

Fruit transpiration could be inferred in the field also through the weight loss of detached fruit (Lang 1990). In that case, the fruit water reservoir is able to sustain the transpiration for several hours (~10) (Li et al., 2002) despite the lack of the xylem/phloem supply by the parent plant. This valuable method is also used to assess in/out fluxes of fruit. Briefly, it is based on fruit diameter (and volume) changes occurring in fruits subjected to a sequence of manipulation as concern their vascular connections (Lang 1990).

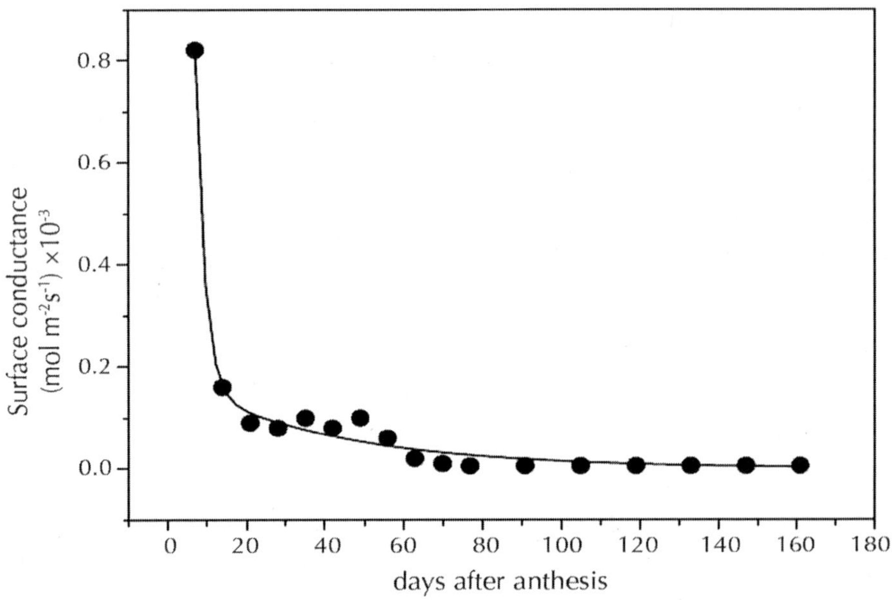

Figure. 1. Surface conductance to water vapour measured in detached kiwifruit. (Redrawn from Smith et al., 1995).

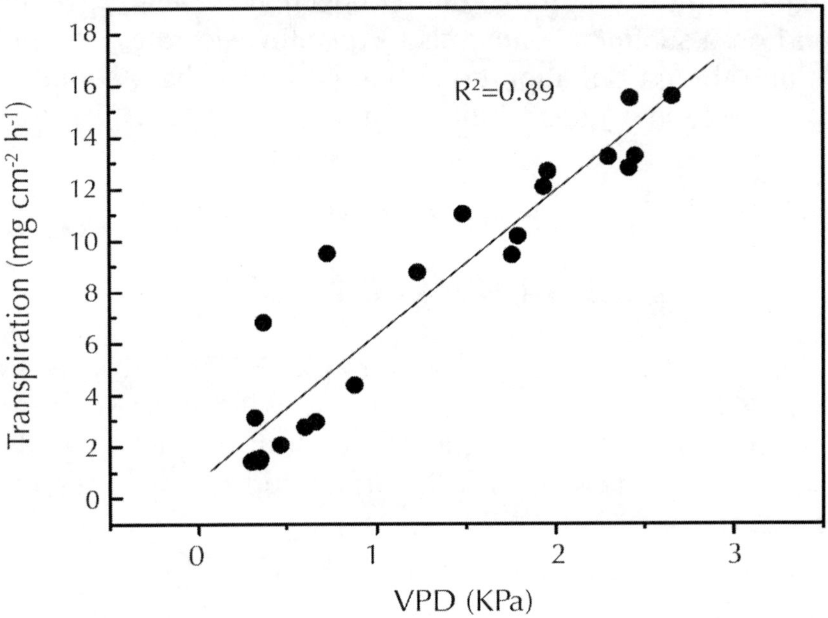

Figure. 2. Relationship between vapour pressure deficit (VPD) and fruit transpiration measured in detached kiwifruit berry in 3-week-old fruits. (Redrawn from Montanaro et al., in press).

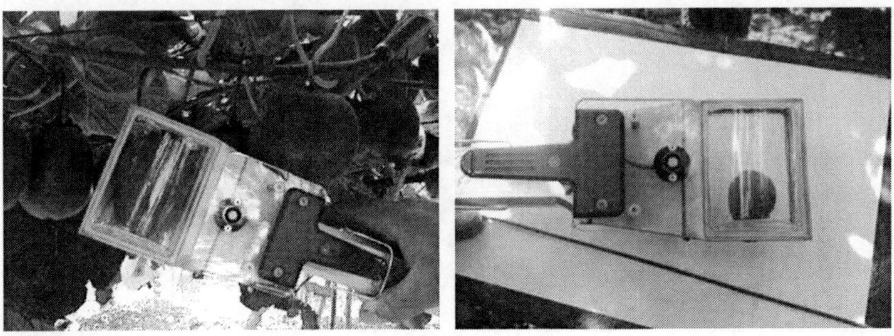

Photo 1. Measurement of fruit transpiration in attached kiwifruit (left) and detached apricot (right) using a portable gas-exchange equipment (ADC-LCA4, ADC BioScientific Ltd, Hoddesdon, England).

The diameter variations of fruit should be accurately determined using specific highly accurate gauge (Lang 1990). That method requires the measurement of diameter of "intact" (with normal vascular connections), "girdled" (with the phloem connection severed) and "detached" fruit (with all vascular connections severed). The method is based on a subtractive analysis to quantify separately the phloem, xylem and the transpiration contribution, under the assumption that volume growth integrates a fruit in/out fluxes.

DAILY AND SEASONAL CHANGES IN FRUIT WATER LOSS

The absolute value of fruit transpiration is inevitably related to the whole fruit surface area. Regular growing fruits show a rapid expansion of their surface area early after fruit set due to cell division (Fig. 3). Later in the season fruit surface further increases but at a lesser extent.

Specific fruit transpiration varies considerably over a 24-hour period taking relatively high values in the middle part of the day but reducing to almost nothing at night (Montanaro et al., in press). Figure 4 shows the daily variation of fruit water loss in two species (peach and kiwifruit) which is presumably due to diurnal variation in key variables of the fruit's aerial microenvironment (e.g. temperature, relative humidity). A noteworthy fact is that despite fruit transpiration was assessed with different method (weigh loss for peach, gas exchange for kiwifruit) it clearly shows a similar behaviour and a consistent absolute value.

Seasonal fruit transpiration usually shows a maximum rate just after fruit set, thereafter it declines quite promptly reaching a minimum at the half of the whole development stage (Fig. 5). In apricot fruit (cv Tyrithos) that maximum has been recorded to be close to 0.55 mmol m^{-2} s^{-1} in very young fruits (i.e. 6 days after fruit-set) (Fig. 5). During the following 4 weeks, transpiration rapidly decreased to approx. 0.35 mmol m^{-2} s^{-1} accounting for about 80% of the whole transpiration decline. Thereafter, transpiration again decreased but more slowly reaching 0.30 mmol m^{-2} s^{-1}.

In kiwifruit, the rate of fruit water loss is similar even the absolute value significantly higher than stone fruits (apricot/peach). Fruit transpiration per unit of fruit area had a maximum value of 2.3 mmol m^{-2} s^{-1} in the first days after fruit set (Fig. 5). During the following 2 weeks transpiration rapidly decreased by 65% compared to the initial values. In the subsequent 30 days it reached very low values, approximately 10% of the initial rate. The substantial reduction in transpiration of fruit by day 50-60 after fruit set occurs at the same time as the permanent disfunction of the fruit vascular system as reported by Dichio et al. (2003).

There are evidences that the minimum rate of transpiration reached by fruit is species specific. The minimum transpiration rate recorded at the end of the grow period in Prunus armeniaca, was relatively high at 55% of the early value according to data on peach (Li et al., 2002). The transpiration profile is also in agreement with findings in grapevine and kiwifruit even though the extent of decline differed (Boselli et al., 1998; Montanaro et al., 2006). In these species, the lowest transpiration rate was about 10% (kiwifruit) and 20% (grapevine) of the initial values. In kiwifruit the large decline of fruit transpiration has been associated with the evolution of a suberised outer cell layers, and death of the outer cells associated with wax biosynthesis (Celano et al., 2009). It is likely that such changes are not so marked in the case of apricot, leading the lowest transpiration rate to remain within about ½ of the initial values.

Changes in Skin Properties

Seasonal changes in fruit-skin conductance are probably responsible for the decline in transpiration during the season showed in Figure 5. Xiloyannis et al. (2001) the decrease in fruit transpiration in kiwifruit has been associated with the collapse of the surface tissues of the fruit and the development of a suberized periderm. That is, transversal freeze fractures of frozen-hydrated kiwifruit epidermis proved that the external layer of epidermal cells collapsed during fruit development (Photo 2), producing a bearing-like protection for the fully hydrated parenchymatic cells positioned below them. These epidermis structural changes are associated with the heavy reduction in transpiration rate recorded during fruit growth (Xiloyannis et al., 2001).

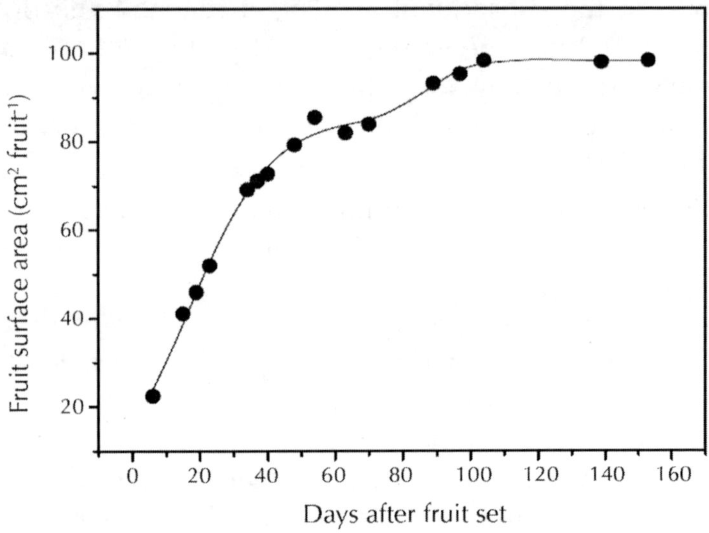

Figure. 3. Seasonal trend of fruit surface area observed in kiwifruit (cv Hayward).

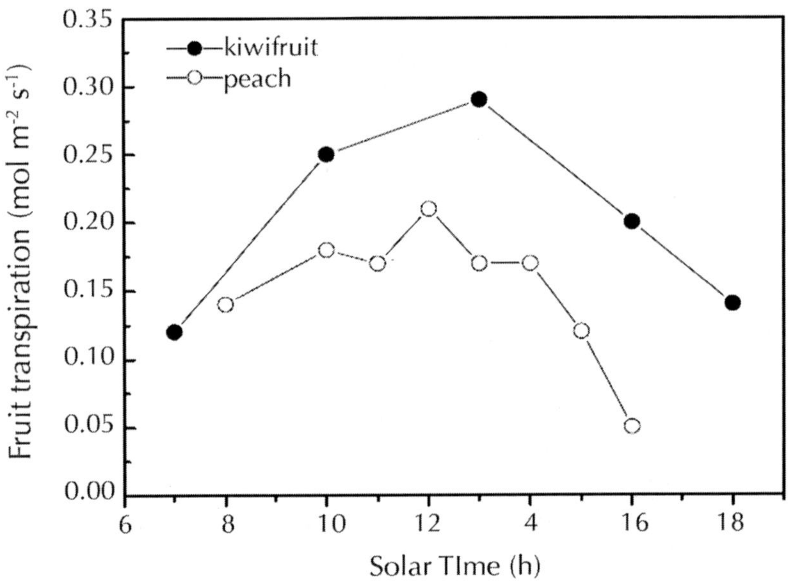

Figure. 4. Daily oscillation of fruit water loss measured 1-week after fruit-set in attached kiwifruit (cv Hayward) and detached peach fruit (cv Dixired). (Redrawn from Li et al., 2002 and Montanaro et al., 2006).

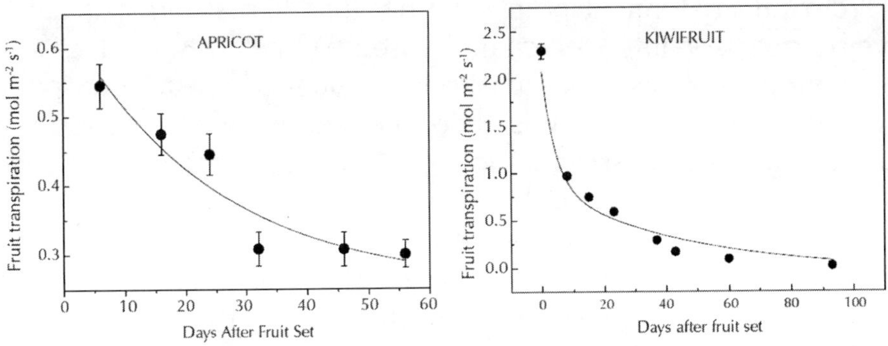

Figure. 5. Seasonal trend of specific fruit transpiration rate measured midday in apricot (left) and kiwifruit (right) as measured in detached and attached fruit, respectively. (Redrawn from Montanaro et al., 2006 and 2010).

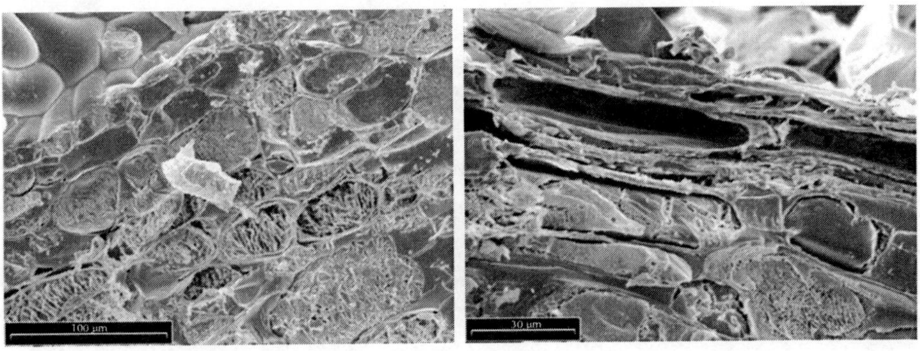

Photo 2. Low Temperature SEM images. Transversal freeze fractures of frozen-hydrated kiwyfruit epidermis sampled at the 4th (left) and at the 17th (right) week from fruit set. Note the collapse of the external layers of epidermal cells during the development of the fruit, which produces a bearing-like protection for the parenchimatic cells positioned below them. (From Xiloyannis et al., 2001).

Which role for hairs?

In some species, fruits have the skin with dermal structures (hairs) which inevitably are involved in skin conductance properties and in turn affect fruit transpiration. This is the case with some kiwifruit species (e.g. Actinidia deliciosa).

From an anatomical point of view the dermal hairs of kiwifruit have been reasonably well characterised (White, 1986) but, as yet no information exists as to their possible influence on the transpiration physiology of the fruit. It is therefore worthwhile at this stage to introduce to the discussion some relevant aspects of their development and physiology. As previously reported in Xiloyannis et al., (2008), the juvenile kiwifruit already has a full set of hairs (trichomes) at bloom and no new hairs develop from this time on. The hairs are also fully formed and are typically 2.5 to 3.0 mm long and their bases are about 0.1 mm in diameter giving them an aspect ratio (length/basal diameter) of about 25. In the very young fruit, the hairs are packed together so closely that their bases are almost touching. Subsequent surface expansion of the fruit moves the hairs apart so that they become increasingly isolated from one another during the growing season.

Hairs are of quite simple structure (Fig. 6) and contain no vascular tissue. This requires that any evaporative water loss occurring from the more apical cells must be made good by diffusion of water up from the skin through the more basal cells. This diffusion pathway (it is likely that it will be predominantly through the cell wall matrix, not through the protoplasts) is probably sufficient to maintain hair vitality early on when the hairs' close proximity to one another means that they will offer a degree of mutual protection from water loss. However, as they move apart with fruit growth, this protection will become less and less, and so the acropetal diffusion pathway will probably become limiting at some stage. This understanding fits with our observation that cell death (drying) is first evident in the most apical hair cell and this then progresses downwards (basipetally) through the hair eventually reaching the most basal ones last (Xiloyannis et al., 2008).

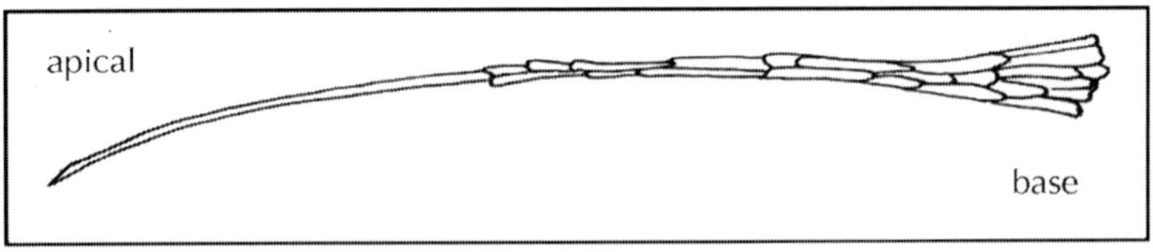

Figure. 6. Drawing of a dermal hair from cv. 'Hayward' kiwifruit. Cell death occurs basipetally, from left to right in this drawing (Redrawn from Xiloyannis et al., 2008).

A simple calculation based upon hair dimensions (the area of a hair ≅ π × base radius × hair length) and hair density (hairs mm^{-2} of fruit surface) (data not shown) proves that, in a very young kiwifruit, the still-living hairs are sufficiently numerous as to increase the 'living' surface area of the fruit by a factor of around 40-times, compared with a hairless fruit of the same dimensions. This massive amplification of fruit surface area is almost certainly responsible for the very high surface conductance of a young kiwifruit. The amplification factor steadily decreases as the fruit surface area increases with growth (data not shown), but the hair dimensions and hair numbers per fruit remain the same (i.e. the hair density decreases). The proportion of living hairs (hairs with some live hair cells) falls from close to 100% at the time of fruit set to about 50% by 30 days after full bloom and by 60 days after full bloom all the cells of all hairs are dead (Fig. 7). We interpret the rather striking skin surface conductance results of Smith et al., (1995) and our own fruit transpiration results, not in terms of the development of a more complex dermal layer (although this too could be involved), but predominantly in terms of the basipetal progression of cell death in the dermal hairs. During this short period of early fruit growth, as hair cell death progresses, the 'hair effect' will change from one in which they increase water loss from the fruit skin to one in which they decrease it. The fully dead (dry) hairs will serve to reduce fruit water loss by entrapping a thick (c. 3 mm) boundary layer of still, moist air around the fruit. This effect is much the same as that of the layer of fur that reduces the rate of heat loss from the skin of mammals. Moreover, this understanding is consistent with the behaviour, elsewhere reported (Dichio et al., 2007) that, under the influence of an increasing forced air stream, the kiwifruit's boundary layer resistance breaks down progressively and fruit transpiration increases very significantly. This in turn impacts the import of certain minerals (see below).

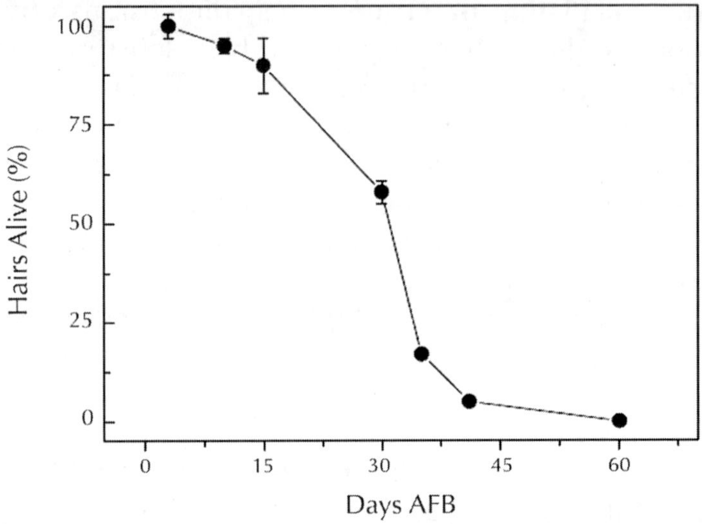

Figure. 7. Hair viability (% of hairs in the sample containing some live cells). Fruit were grown under normal commercial conditions in New Zealand and were not subjected to any special treatment. Each point is the average (± SE) of 100 observations. (Redrawnm from Dichio et al., 2003).

The within-fruit Resistances

Late season occurrence of anatomical changes of fruit epidermis produces a reduction of surface conductance (Smith et al., 1995; Celano et al., 2009) contributing to the transpiration decline previously observed during fruit development (Montanaro et al., 2006), this might strongly reduce the driving force of xylem inflow. Reduction of incoming xylem stream would be also the result of various co-occurring phenomena such as disruptions (or occlusions) localized either in the (tomato) pedicel (Van Ieperen et al., 2003) or in the (apple, kiwifruit) fruit tissue (Dichio et al., 2003; Dražeta et al., 2004). In fruit of Vitis sp the reduction of hydrostatic pressure gradients and the high phloem flows at the end of the season have been indicted to repress xylem inflow (Bondada et al., 2005; Keller et al., 2006; Choat et al., 2009). In some crops (e.g. apple, tomato and grape, kiwifruit) it has been argued that high hydraulic resistance inside fruits may be involved in phloem unloading processes and to protect the fruit from excessive backflow (Keller et al., 2006 and references therein; Morandi et al., 2010a).

Figure 8 shows the seasonal changes in the hydraulic conductance of fruit (berry + stalk) of kiwifruit. Within the first 35 days after full bloom, hydraulic conductance increased rapidly. This is presumably due to the differentiation of new xylem vessels as a result of a post bloom rise in the activity of the intrafascicular vascular cambium, probably stimulated by seed set (Dražeta et al., 2004). Conductance rose to about 0.22 $cm^3\ MPa^{-1}\ s^{-1} \times 10^{-3}$. For about 20 days, conductance remained relatively stable but then decreased progressively after day 60 after full bloom, according to Dichio et al. (2003). In the late season, hydraulic conductance (measured using a pressure bomb) was very low (0.05 -0.1 $cm^3\ MPa^{-1}\ s^{-1} \times 10^{-3}$) (Fig. 8). This result is in conflict with the data emerging from our dye studies (Dichio et al., 2003), which show a much earlier cessation of flow. This discrepancy will be addressed more thoroughly in a separate study and may have to do with the rather high axial pressure gradient applied to the xylem vessels resulting from the application of a bomb pressure of 1.3 MPa (this high bomb pressure is a requirement of our method in order to obtain measurable volumes of exuded sap). The problem is that the pressure is applied over a relatively short distance of about 70 mm between the fruit flesh and the proximal end of the fruitstalk – this is an unnaturally high gradient for the xylem that would normally experience axial gradients some 3-orders of magnitude less. An alternative explanation of the discrepancy (Bondada et al., 2005) could be a reduction in the available pressure gradients (the driving force of xylem sap flow), but this explanation requires confirmation through further experimentation and analysis. The picture emerging emphasises that physical alterations along the fruit xylem pathway do account for some of the reduction in xylem water inflow to the fruit in the late season and consequently the lowered rates of Ca import.

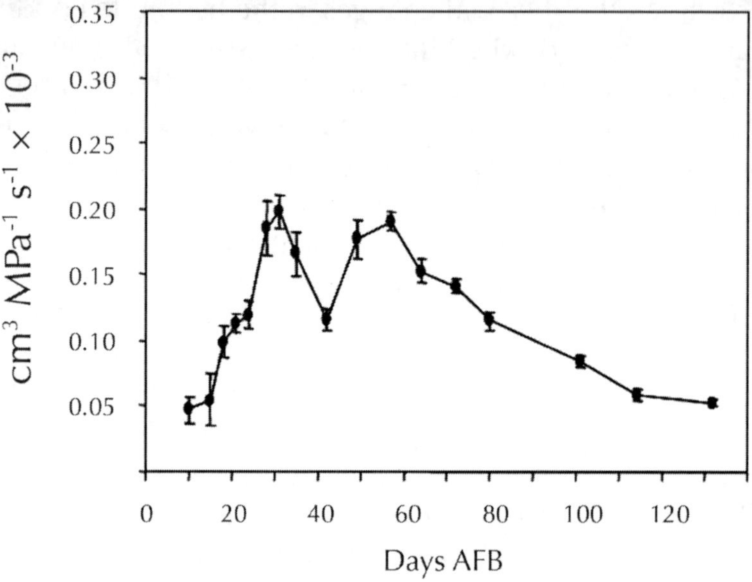

Figure. 8. Hydraulic conductance of kiwifruit fruit (berry + peduncle) measured using a pressure bomb method. Each point is the average of 12 fruit. (Redrawn from Xiloyannis et al., 2008).

Calculating diurnal fruit Transpiration

The diurnal fruit transpiration could be calculated through the integration of the fitting curve of the daily measurements. By plotting the single diurnal values, the seasonal trend of the total fruit transpiration could be estimated. For example, in kiwifruit it has been observed a considerable increase in daily fruit transpiration during the early first 25 days after fruit set when a value of approx. 2 g of water per fruit per day was reached (Fig. 9). Subsequently, fruit transpiration decreased. In fact, at 43 day after fruit set, the reduction in fruit transpiration per day with respect to the peak value observed was 65%, later in the season (93 day) fruit transpiration was almost zero (Fig. 9). The increase in total fruit transpiration in the early part of fruit growth apparently contrasts with the reduction of the rate of fruit water loss (Fig. 5) usually observed in that early stage. The rapid increase of fruit area occurring early of fruit development (Fig. 3) helps to explain such a raise of total amount of water lost by fruit. Measurements of transpiration flux (rate and seasonal trend) of kiwifruit are comparable with findings in stone fruit (detached peach, Li and co-workers 2002) measured in field by weight loss method.

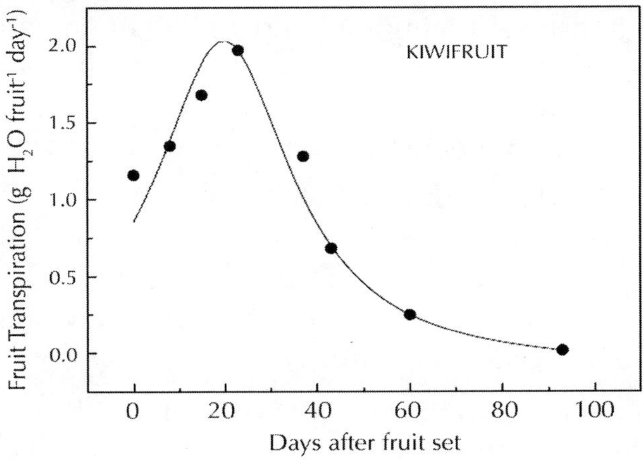

Figure. 9. Seasonal fruit transpiration calculated in kiwifruit (cv Hayward). Each point is the integral of 5 measurements recorded from 7 pm to 6 am.

SIGNIFICANCE OF TRANSPIRATION ON DRY MATTER AND MINERALS ACCUMULATION

Fruit transpiration represents the key driver for accumulation of xylem-born minerals, while the growth of fruit depends on phloem stream. However, recently a role of fruit water loss for the phloem unloading into fruit has been proposed. Water losses induce a decrease in fruit water potential (more negative) via an increase in the osmotic concentration and a decrease in the turgor pressure (Morandi et al., 2010b). In this physiological model, as more water is lost by a fruit, more water can be drawn from the phloem and the xylem streams into the fruit itself. However, other regulatory factors, like environmental conditions, related to time of the day, and stem/fruit water potential gradients, may affect fruit inflows. During the central part of the day, the xylem import to the fruit is low, likely due to the high amount of water directed to transpiring leaves, which reach lower water potentials (Morandi et al., 2010b). Xylem inflows cannot balance the high transpiration water losses, with the consequence that fruit shrink (Morandi et al., 2007), increase their concentration and decrease their turgor pressure. This may facilitate translocation and bulk flow phloem unloading into fruit tissues relating fruit transpiration to phloem import, hence transpiration may be viewed as important in determining fruit daily imports of water and dry matter (Morandi et al., 2010b).

Based on this idea, reduction of fruit transpiration would reduce accumulation of dry matter. The Figure 10 shows the effect of reduced transpiration (obtained through bagging which saturates the environment surrounding the fruit leading the relative humidity close to 100%) in apricot and peach fruits.

Optimal mineral nutrition represents a prerequisite for achievement and preservation of high quality plant product. Particular attention should be paid to calcium nutrition due to its involvement in determining tissue mechanical strength and tolerance to biotic and abiotic stresses (Hirschi, 2004). Calcium is a phloem immobile element (Bukovac and Wittwer, 1957), this implies that the amount of Ca reaching a fruit (or a leaf) is almost entirely dependent on supply trough the xylem. Xylem transportation of Ca depends on plant and environmental variables (and their interaction) which often are unfavourable and reduce Ca accumulation in fruit (White and Broadley, 2003).

In kiwifruit, it has been proved the positive effect of windspeed on Ca accumulation (Dichio et al., 2007. That is, for low average windspeeds (0–1.5 m/s), Ca remained almost constant, the level ranging from 18 to 24 mg/fruit. As windspeeds increased beyond this range, the level of mineral rose significantly. Applying a simple cubic model to the data ($y = a + bx^3$) the r2 value obtained for this rise were 0.72 (Fig. 11). For the highest windspeeds of 3.3 m/s (about 12 km/h) the levels of Ca was almost double that for still air (see Fig. 11). In contrast with Ca, the accumulation of fruit K (mass per fruit) showed no evidence at all of any rise with windspeed. Instead K levels were about 300 mg/ fruit (ranging from 212 to 380 mg/fruit) and this level was maintained regardless of windspeed. Applying the same cubic model to the K results gave an r2 value of only 0.13.

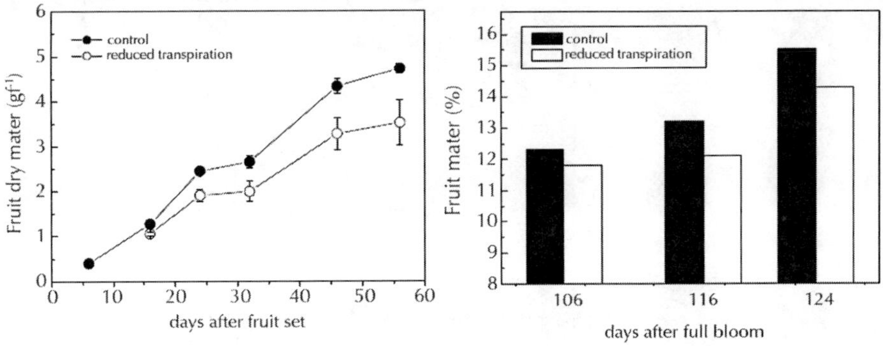

Figure. 10. Dry matter accumulation in fruit of apricot (left) and peach (right) measured in transpiring fruit (control) and under reduced transpiration. (Redrawn from Montanaro et al., 2010 and Morandi et al., 2010b).

Plant water status and transpiration play a key role in supplying the various plant tissue with Ca, particularly those with a low transpiration rate such as fruit (Bangerth, 1979). Mechanisms by which Ca is accumulated into fruit have been searched in a number of fleshy and stone fruit (e.g. apple, tomato, kiwifruit, apricot) (Ferguson and Watkins, 1989; Montanaro et al., 2006 and 2010; Liebisch et al., 2009) with fewer attempts to separate transpirational mechanisms apart from others (e.g. hormonal, metabolic). Recently, a study was undertaken to (i) investigate the relationship between accumulation of Ca and transpiration in developing apricot fruit, and (ii) determine the prominence of transpirational flux upon the whole transpiration-independent transportation processes on Ca nutrition (Montanaro et al., 2010). The hypothesis behind this work was that if the fruit transpiration is the determining factor of the accumulation of Ca (phloem-immobile element) then the import of Ca would be suppressed by restriction of fruit water loss, while the import of phloem-mobile nutrients (i.e. K and Mg) would not be. To test this hypothesis, the seasonal changes of fruit transpiration and Ca, K and Mg concentrations and accumulation were assessed in fruits left to naturally transpire (control) or under restricted transpiration. Restriction of transpiration was obtained through bag application (Photo 3).

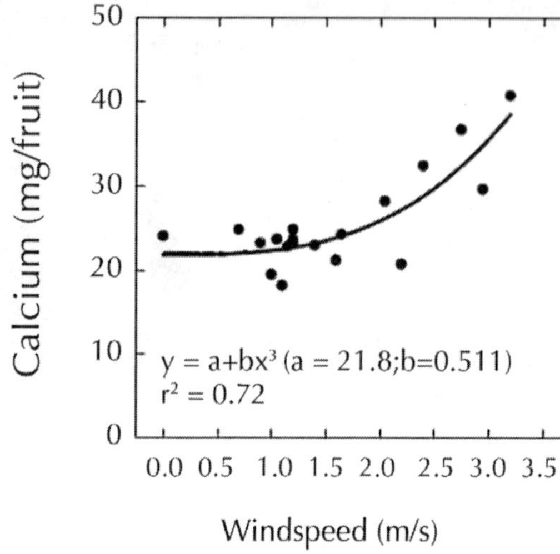

Figure. 11. Whole fruit mineral content for calcium as a function of the mean windspeed to which individual fruit were exposed throughout the growing season. Each point represents a single fruit. (Redrawn from Dichio et al., 2007).

Photo 3. A view of zip-lock bags just mounted on fruit within the canopy of apricot trees (left). After a couple of hours in the microenvironment surrounding the fruit the water vapour condensed (right), hence the relative humidity assumed at saturation point.

Bagging treatments have been widely used to search for fruit response to certain growth condition in terms of skin colour, mineral composition, size, maturity, etc. (Hofman et al.,

1997; Amarante et al., 2002). In this study, bags were installed to keep the relative humidity of the air surrounding fruit close to saturation point in order to test whether the limitation of water loss by fruit suppresses the import of some nutrients.

In non-transpiring fruits Ca concentration was significantly affected by restriction of transpiration (it was approximately 30% of that of control fruit), while phloem-mobile elements (K, Mg) were not (Fig. 12).

The evidence that restriction of transpiration did not exert any significant effect on K and Mg concentrations, suggests that the lowest amount of Mg and K accumulated in non- transpiring (bagged) fruit is attributable to the 25% reduction of DM import rather than to the reduced concentration (Fig. 12). At fruit scale, preservation of saturating air humidity significantly lowered the import of Ca due to reduced concentration and reduced DM import according to findings in apple (Tromp and Van Vuure, 1993).

In this study, although plants grew on a Ca-rich soil, the non-transpiring fruits were Ca- deficient compare to control fruits (Fig. 12) and showed the typical visual symptoms of chlorosis. Development of punctual signs of Ca-related disorders have for a long time been associated with mechanisms involved in the movement of Ca within plant parts (Simon, 1977) can conceivably to be explained by differences in transpiration.

The evidence that even though transpiration was negligible, some calcium entered the fruit supports the idea that transpiration is not the only factor governing the movement of Ca through its strong function as driving force of the xylem stream (Bangerth, 1979). Partitioning of Ca amidst plant organs is imputable also to metabolic demand and chemical aspects of the conductive tissues (e.g. adsorption and desorption processes occurring at exchange sites along the walls of the xylem pathway) (McLaughlin and Wimmer, 1999). Calcium delivery to fruit has also been associated to hormonal activities. For example, a mutual relationship between polar basipetal auxin transport and acropetal Ca transport has been reported for tomato, apple and avocado (Stahly and Benson, 1970; Bangerth, 1976; Banuelos et al. 1987; Cutting and Bower, 1989). However, there is still limited information to discriminate (and quantify) the effect of these transpiration- independent stimuli in Ca nutrition. Interpretation of the ratio between concentrations detected in non-transpiring and control fruit would be valuable for assessing of the prominence of transpiration on Ca nutrition.

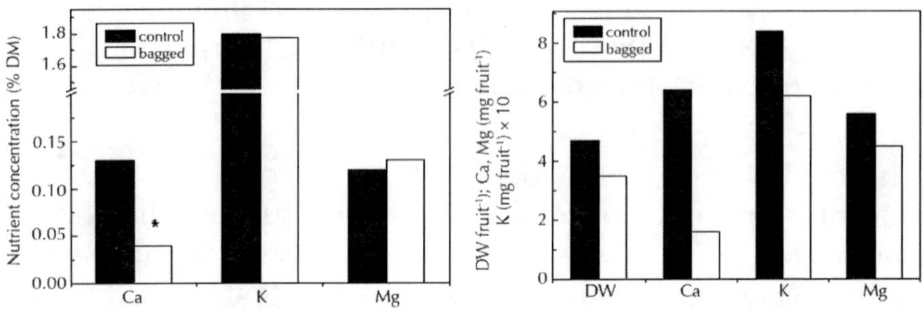

Figure. 12. Concentrations of calcium (Ca), potassium (K) and magnesium (Mg) (left) and amounts per fruit of dry matter (DW), Ca, K and Mg measured at harvest in naturally transpiring apricot fruit (control) and non-transpiring (bagged) fruit. (Redrawn from Montanaro et al., 2010). (For the left figure, * indicates a significant difference at $p < 0.05$, Student's t-test between treatments).

In the case of Ca, that ratio was 0.45 about 50 days after bags were positioned on a whole- bagged period basis. Hence, the fruit transpiration accounted for 55% of the Ca concentration gained by fruit. By contrast, that ratio for K and Mg was not severely affected by the restriction of transpiration, so that the ratio was on average 0.97 and 0.91 for K and Mg respectively, indicating a weak effect of restricted transpiration. For K and Mg the transpiration-independent factors operated as well, however their high phloem mobility does not allow for their unmasking.

Calcium movement in plants has been extensively studied in relation to factors such as transpiration flux, hormonal activities, nutrient demand and chemical properties of the conductive system (Bangerth, 1979; Banuelos et al., 1987; Cutting and Bower, 1989; McLaughlin and Wimmer, 1999), however information is still limited to discriminate the relative prominence of each individual factor. Our data were insufficient to indicate the transpiration-independent mechanism(s) by which 45% of the Ca concentration accrued under high humidity condition, but the hormones were possibly involved. For example, in the case of tomato fruit, the polar basipetal auxin transport has been indicted to promote about 15% of Ca concentration in fruit kept at saturating RH (Banuelos et al., 1987). In addition, in kiwifruit a correlation between Ca accumulation and auxin has been reported (Fig. 13).

Calcium accumulation in fruit could be affected also by other factors such as the position of fruiting sites within the canopy. For example, apple fruit in the upper parts of the canopy tend to have lower Ca concentrations than those in the lower parts, the age (and others features) of leaves of fruiting branches may also affect Ca accumulation in fruit (Volz et al., 1996). It as been observed in apple fruit that reducing the leaf:fruit ratios at specific fruiting sites resulted in lower Ca concentrations, with the activity (presumably transpiration) of spur leaves being especially important (Volz et al., 1996; Lang and Volz, 1998).

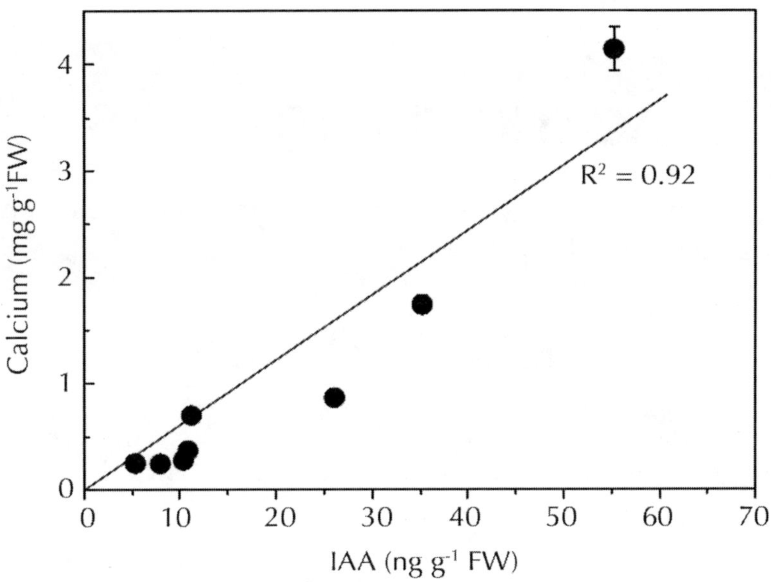

Figure. 13. Correlation between calcium and auxin (IAA) detected in approx. 8-week-old kiwifruit berry (Redrawn from Sorce et al., 2011).

In kiwifruit, relationships exist between fruiting position, leaf:fruit ratios and fruit mineral concentrations. Thorp et al. (2003) demonstrated that a relatively small shift in leaf:fruit ratio was sufficient to affect fruit Ca concentrations, but only in regions of the vine with relatively low fruit mineral levels and low leaf areas. In the same study, it has been highlighted the role of cane size on that relationship. Large diameter canes have axillary shoots with relatively high leaf areas compared with shoots on small diameter canes, especially at basal positions on the cane (Thorp et al., 2003).

Mechanisms behind the effect of leaves on Ca accumulation in fruit need to be clarify. It should be hypothesised that Ca accumulation in a plant organ is proportional to the sap flow unloaded by the organ and to the relative Ca concentration in sap:

$$Ca = [Ca] \times SAP_{Flow}$$

Hence, transpiration of leaves may helps to drive more sap which carries nutrients. To explain the effect of fruit position and shoot type on Ca accumulation, it could be evoked that the relative rate of Ca use along the transport pathway may affect [Ca]. Hence, fruits located at the tip of a shoot tend to receive a Ca-poor sap compare to fruits at the base.

ACKNOWLEDGMENTS

This work was funded by Ministero dell'Istruzione, dell'Università e della Ricerca - Prin2009 grant.

REFERENCES

1. Amarante, C., Banks, N. H., Max, S., 2002. Effect of preharvest bagging on fruit quality and postharvest physiology of pears (Pyrus communis). N. Z. Crop Hort. Sci. 30, 99-107.
2. Bangerth, F. 1976. A role for auxin and auxin transport inhibitors on the Ca content of artificially induced parthenocarpic fruits. Physiol. Plant. 37, 191-194.
3. Bangerth, F. 1979. Calcium-related physiological disorders of plants. Annu. Rev. Phytopathol. 17, 97-122.
4. Banuelos, G. S., Bangerth, F., Marschner, H. 1987. Relationship between polar basipetal auxin transport and acropetal Ca transport into tomato fruits. Physiol. Plant. 71, 321-327.

5. Bondada, B.R., Matthews, M.A. and Shackel, K.A., 2005. Functional xylem in the post- veraison grape berry. J. Experim. Bot. 56:2949–2957.
6. Boselli, M., Di Vaio, C., Pica, B. 1998. Effect of soil moisture and transpiration on mineral content in leaves and berries of Cabernet Souvignon grapevine. J. Plant Nutr. 21(6), 1163-1178.
7. Bukovac, M.J., Wittwer, S.H. 1957. Absorption and mobility of foliar applied nutrients. Plant Physiol. 32, 428-435.
8. Celano, G., Minnocci, A., Sebastiani L., D'Auria, M., Xiloyannis, C. (2009): Changes in the structure of the skin of kiwifruit in relation to water loss. J. Hort. Sci. Biotechnol. 84, 41-46.
9. Choat, B., Gambetta, G.A., Shackel, K.A., Matthews, M.A., 2009. Vascular function in grape berries across development and its relevance to apparent hydraulic isolation. Plant Physiol., 151, 1677–1687.
10. Cline, J.A., Hanson, E.J., 1992. Relative humidity around apple fruit influences its accumulation of calcium. J. Amer Soc Hort Sci 117(4), 542-546.
11. Cutting, J. G. M., Bower, J. P. 1989. The relationship between basipetal auxin transport an calcium allocation in vegetative and reproductive flushes in avocado. Sci. Hort. 41, 27-34.
12. Dichio B., Remoroni D., Lang A., 2007. Calcium accumulation in fruit of kiwifruit grown under different windspeed conditions. Acta Hort., 753:509-514.
13. Dichio, B., Remorini, D., Lang, A. 2003. Developmental changes in xylem functionality in kiwifruit fruit: implications for fruit calcium accumulation, Acta Hort. 610:191- 195.
14. Dražeta, L., Lang, A., Hall, A.J., Volz, R.K., Jameson, P.E., 2004. Causes and effects of changes in xylem functionality in apple fruit. Ann. Bot. 93, 275-282.
15. Faust, M., Shear, C.B., Smith, C.B. 1968. Investigation of corking disorders of apples II. Chemical composition of affected tissues. Proc Amer Soc Hort Sci 92:82-88.
16. Ferguson, I.B., Thorp, T.G., Barnett, A.M., Boyd, L.M., Triggs, C.M. 2003. Inorganic nutrient concentrations and physiological pitting in 'Hayward' kiwifruit. J Hort Sci Biotech 78:497-504.

17. Fiscus, E.L., Klute, A., Kaufmann, M.R. 1983. An interpretation of some whole plant water transport phenomena. Plant Physiol. 71: 810-817.
18. Goudriaan, J., van Laar, H. H. 1994. Modelling Potential Crop Growth Processes. Kluwer Academic Publ., Dordrecht – Boston – London, p. 238.
19. Hirschi, K. D. 2004. The calcium conundrum. Both versatile nutrient and specific signal. Plant Physiol. 136, 2438–2442.
20. Hofman, P. J., Smith, L. G., Joyce, D. C., Johnson, G. I., Meiburg, G. F. 1997. Bagging of mango (Mangifera indica cv. "Keitt") fruit influences fruit quality and mineral composition. Postharvest Biol. Technol. 12, 83-91.
21. Jarvis, PPG., 1985. Coupling of transpiration to the atmosphere in horticultural crops: The Omega factor. Acta Hort. 171, 187-206
22. Keller, M., Smith, J.P., Bondada, B.R., 2006. Ripening grape berries remain hydraulically connected to the shoot. J. Experim. Bot. 57(11), 2577–2587.
23. Lang, A., Volz, R.K., 1998. Spur leaves increase calcium in young apples by promoting xylem inflow and outflow. J. Amer. Soc. Hort. Sci. 123, 956-960.
24. Lang, A. 1990. Xylem, phloem and transpiration flows in developing apple fruits. J. Exp. Bot.41, 645-651.
25. Li, S. H., Génard, M., Bussi, C., Lescourret, F., Laurent, R., Basset, J., Habib, R. 2002.
26. Preliminary studies on transpiration of peaches and nectarines. Eur. J. Hort. Sci. 67, 39-43.
27. Liebisch, F., Max, J. F. J., Heine, G., Horst W. G. 2009. Blossom-end rot and fruit cracking of tomato grown in net-covered greenhouses in Central Thailand can partly be corrected by calcium and boron sprays. J. Plant Nutr. Soil Sci. 172, 140 – 150.
28. McLaughlin, S. B., Wimmer, R. 1999. Ca physiology and terrestrial ecosystem processes. New Phytol. 142, 373-417.
29. Montanaro, G., Dichio, B., Xiloyannis, C., 2010. Significance of fruit transpiration on calcium nutrition in developing apricot fruit. Journal of Plant Nutrition Soil Science, 173: 618–622. doi: 10.1002/jpln.200900376

30. Montanaro, G., Dichio, B., Xiloyannis, C., Celano, G. 2006. Light influences transpiration and calcium accumulation in fruit of kiwifruit plants (Actinidia deliciosa var. deliciosa). Plant Sci. 170, 520-527.
31. Montanaro, G., Dichio, B., Xiloyannis, C., Lang, A., in press. Preliminary evaluation of the transpiration response of young actinidia fruit to the weather. Proceedings of the VII International Symposium on Kiwifruit, ISHS. 12-17 September 2010, Faenza – Italy.
32. Morandi, B, Rieger, M.W., Corelli Grappadelli L., 2007. Vascular flows and transpiration affect peach (Prunus persica Batsch) fruit daily growth. J Exp Bot 58: 3941–7.
33. Morandi, B., , Manfrini, L., Losciale, P., Zibordi, M., Corelli-Grappadelli, L., 2010a. Changes in vascular and transpiration flows affect the seasonal and daily growth of kiwifruit
34. (Actinidia deliciosa) berry. Ann. Bot., 105, 913–923.
35. Morandi, B., Manfrini, L., Losciale, P., Zibordi, M., Corelli-Grappadelli L. 2010b. The positive effect of skin transpiration in peach fruit growth. J. Plant Physiol., 167:1033-1037.
36. Nobel, P.S. 2005. Physiochemical and environmental plant physiology - 3rd ed. Burlington: Elsevier Academic Press.
37. Simon, E. W. 1977. The symptoms of calcium deficiency in plants. New Phytol. 80, 1-15. Smith, K.U., Clages, T.G.A., Green, E.F., Walton, Changes in abscissic acid concentration, surface conductance, and water content of developing kiwifruit, Sci. Hortic. 61 (1995) 13-27.
38. Sorce, C., Lombardi, L., Remorini, D., Montanaro, G. 2011. Occurrence of natural auxin and accumulation of calcium during early fruit development in kiwifruit. Australian Journal of Crop Science, 5(7): 895-898.
39. Stahly, E. A., Benson, N. R. 1970. Calcium levels of "Golden Delicious" apples sprayed with 2, 3, 5-Triidrobenzoic acid. J. Am. Hort. Sci. 95, 726-727.
40. Thorp, T.G., Ferguson, I.B., Boyd, L.M., Barnett, A.M., 2003. Fruiting position, mineral concentration and incidence of physiological pitting in "Hayward" kiwifruit. J. Hort. Sci. Biotechnol., 78, 505-511.

41. Tromp, J., Van Vuure, J. 1993. Accumulation of calcium, potassium and magnesium in apple fruits under various conditions of humidity. Physiol. Plant. 89, 149-156.
42. Tzoutzoukou, C. G., Bouranis, D. L. 1997. Effect of preharvest application of calcium on the postharvest physiology of apricot. J. Plant Nutr. 20, 295-309.
43. Volz, R. K., Tustin, D. S. and Ferguson, I. B. 1996. Mineral accumulation in apple fruit as affected by spur leaves. Scientia Horticulturae, 65, 151-161.
44. White, J. 1986. Ontogeny and morphology of ovarian and fruit hairs in kiwifruit. New Zeal J Bot 24:403-414.
45. White, P., Broadley, M. R. 2003. Calcium in plants. Ann. Bot. 92, 487-511.
46. Xiloyannis C., Dichio B., Montanaro G., Lang A., Celano G., Mazzeo M., 2008. Fruit morphological and physiological traits influence calcium transport and accumulation in kiwifruit. Acta Hort., 767:369-378
47. Xiloyannis, C., Celano, G., Montanaro, G. Dichio, B., Sebastiani, L., Minnocci, A. 2001. Water relations, calcium and potassium concentration in fruits and leaves during annual growth in mature kiwifruit plants. Acta Hort. 564: 129-134.

Chapter 6

Rhizospheric Microflora Escalating Aroma Constituents and Yield Attributes in Ocimum Tenuiflorum (L.) cv. CIM-Ayu

Shilpi Khare Saikia and Rakesh Pandey

Microbial Technology and Nematology Department, CSIR-Central Institute of Medicinal and Aromatic Plants, P.O. CIMAP, Lucknow 226015, India

ABSTRACT

The exploration of rhizospheric microbial flora for crop yield enhancement is well established. Rhizospheric microbes influence the plant physiology by imparting several beneficial effects, namely, Nitrogen fixation, increased nutrient uptake, and secondary metabolites production on their host plants. The present study investigates the response of Bacillus megaterium ATCC No. 13525, Pseudomonas fluorescens ATCC No. 14581, and Trichoderma viride MTCC No. 167 in alone and combined treatments for their effect on growth and yield parameters in a commercially important Ocimum tenuiflorum L. cv. CIM-Ayu. The plant is therapeutically important for its essential oil constituents, namely, eugenol, β-caryophyllene, and various monoterpenes. The combination treatments, T7 (B. megaterium + P. fluorescens) and T8 (B. megaterium +P. fluorescens + T. viride), showed maximum enhancement (27.27%) of percentage essential oil as compared to untreated control. Nutrient uptake especially N_2 content was significantly increased (43%) with the treatment T8 (B. megaterium + P. fluorescens + T. viride). Amongst major essential oil constituents, eugenol content was maximally increased by 58.5% as compared to 42.9% (control) indicating a cumulative role of microbial inoculants for crop yield boost-up.

INTRODUCTION

The members of genus Ocimum (Gk. ozo = smell) from family Lamiaceae are medicinally important plants owing to their therapeutic potentials as antiseptics, antioxidants, antistressors, antipyretics, antimicrobials, and insecticidal [1]. Ocimum tenuiflorum is an imperative species from the genus, widely cultivated for its high essential oil yields (0.5 to 0.7%), rich in eugenol, methyl chavicol, and linalool content [2]. The commercial importance of essential oil depends on its constituents and hence methods are required to enhance these valuable phytomolecules.

Rhizospheric microorganisms are recognized as an economic and sustainable input for increasing the productivity of several agricultural, horticultural, forestry, and medicinal crops [3–5]. Numerous microbes such as Pseudomonas, Azospirillum, Azotobacter, Klebsiella, Enterobacter, Alcaligenes, Arthrobacter, Burkholderia, Bacillus, and Serratia have been reported to enhance the plant growth [6, 7]. These microorganisms coexist in the rhizosphere, which is a thin soil layer immediately surrounding plant roots. Rhizospheric microbial wealth is widely acclaimed in agricultural practices for enhancement of crop yield attributes as they share their environment with the host plants and thus exhibit better adaptation [8]. Rhizospheric microbes play significant role/s in improving the growth and yield of host plant [9] by imparting several beneficial effects, namely, N_2 fixation, increased nutrient uptake, siderophores, and secondary metabolite/s production [8, 10]. This interest is linked to environmental concerns for reduced use of chemicals as well as an appreciation for utilization of biological and organics in agriculture.

Therefore, a proper understanding of microbial species associated with the host plants is essential in improving the quality and yield in a desired crop. In this regard, the present experiment was designed to study the influence of selected rhizospheric microbial inoculants (previously proven as effective biocontrol agents), namely, Bacillus megaterium ATCC No. 13525, Pseudomonas fluorescens ATCC No. 14581, andTrichoderma viride MTCC No. 167 under greenhouse conditions on the agronomical attributes of O. tenuiflorum cv. CIM-Ayu. The study also validates the role of microbial inoculants for plant growth promotion and alterations in aromatic oil symphony due to them.

MATERIALS AND METHODS

Plant Materials and Growth Conditions

Seeds of high yielding cultivar of O. tenuiflorum "CIM-Ayu," obtained from the National Gene Bank for Medicinal and Aromatic Plants at the Central Institute of Medicinal and Aromatic Plants (CSIR-CIMAP), Lucknow, India, were surface-sterilized by soaking in 10% (v/v) sodium hypochlorite solution for 5 minutes, washed with distilled water, and soaked for 4 h. After sterilization and soaking, healthy looking uniform sized seeds were sown in content plug plates filled with sterilized soil in a greenhouse under natural light conditions, a daytime temperature of about 28°C and relative humidity of 65–70%. Twenty-one days after sowing, four healthy leafed stage seedlings were transplanted into 7.0 kg soil capacity clay pots containing a mixture of autoclaved soil (76% sand, 8% silt, and 16% clay, pH 7.7) and composted farm manure in 5 : 1 ratio. Soil mineralizable nitrogen, phosphorus, and potassium content were estimated [11–13]. Plants were irrigated manually at alternate days to ensure adequate soil moisture throughout the experimentation.

Microbial Inoculants (Culture and Maintenance)

The rhizospheric microbes Bacillus megaterium (ATCC-13525), Pseudomonas fluorescens (ATCC No. 14581), and T. viride (MTCC-167) are continuously maintained in the Microbial Technology and Nematology Department, CSIR-CIMAP, Lucknow. These microbes have a proven role in biocontrol experiments [14–16] and thus were selected for observing plant growth promotion effects. The fungus T. viride was cultured using sand maize media while the bacterial strains were cultured in Luria broth. The fungal seed culture was incubated at 30 ±1°C for 96 h. After mass-multiplication, the mycelial mat with conidia was homogenized and suspended in 500 mL of 0.1 M phosphate buffer (K_2HPO_4; KH_2PO_4) adjusting the colony forming units (CFU) at 1.2×10^6 per mL. The bacterial cultures were incubated at 28 ± 1°C for 48 h under shaking conditions (200 rpm). After proper multiplication, the cultures were centrifuged at 6000 g for 10 min. The supernatants were discarded and the pellets containing bacterial cells were suspended in 500 mL of 0.9% saline, adjusting the CFU as 2.5×10^8 per mL^{-1} for B. megaterium and 1.8×10^8 for P. fluorescens. Bacterial cell suspensions were inoculated as 10 mL of 10^8 CFU/pot whereas the fungal seed culture was inoculated as 10 mL of 10^6 CFU/pot. Nine microbial treatments were considered (T1 = uninoculated control, T2 = B. megaterium, T3 = P. fluorescens, T4 = T. viride, T5 = B. megaterium + T. viride, T6 = P. fluorescens + T. viride, T7 = B. megaterium + P. fluorescens, and T8 = B. megaterium + P. fluorescens + T. viride). Nitrogenous fertilizers were added to all the treatments according to the soil test. Shoot height, root length, shoots and root weight (fresh and dry), and N, P, and K content were determined after 90 days of microbial inoculation, at the harvest of the crop.

Chemical Composition of Essential Oil (GC-Analysis)

Essential oil from the aerial biomass was hydrodistilled using Clevenger apparatus for 4 h. The oil was dried over anhydrous sodium sulphate (Na_2SO_4) and then analysed for gas chromatography (GC) on an Agilent Perkin Elmer GC Instrument (Model) fitted with FID (flame ionization detector) and electronic integrator with (30 m × 32 mm i.d., 0.25 μm film thickness) fused-silica capillary column. Nitrogen (at 0.4 mL/min) was used as a carrier gas. Samples were injected in the split mode at a ratio of 1:10–1:100. The injector was kept at 250°C and the transfer line at 280°C. The column was maintained at 50°C for 2 min and then programmed to 220°C at 5°C/min and held for 10 min at 300°C. The relative quantities of individual components of essential oil were calculated based on software computed GC peak area percentage without applying correction for FID response factor. The identification of the compounds was performed by comparing their retention indices (RI), determined with reference to a homologous series of n-alkanes (C9–C24, Polyscience Corp.) under identical experimental conditions in both polar and nonpolar columns, coinjection with standards. The relative quantities of the individual components were calculated based on computer calculated GC peak areas without correction for flame ionization detection response factors.

Nutrient Uptake Analysis

Soil mineralizable nitrogen, phosphorus, and potassium content were estimated in various treatments for assessing growth promotion effects. Initial soil samples were analyzed for available N ($kg\,ha^{-1}$), available P (Olsen's P_2O_5, $kg\,ha^{-1}$), and available K (exchangeable K_2O, $kg\,ha^{-1}$) following Jackson [17]. The estimations were made through FOSS application notes AN5222, ENISO 11732:1997, and AN5249 ISO 6878: 1996 with slight modifications in flame photometer 128, Cistronic model.

Statistical Analysis

The plant treatment was conducted in a random complete block design with three replicates per treatment. The data were subjected to analysis of variance (ANOVA) by using Duncan's multiple test, and treatment means were compared using least significant difference at P=0.05. Least significant difference (LSD) was calculated at 5% probability level () for comparing the significance of difference between any two treatment means [18]. The statistical analyses were performed using ASSISTAT 2012 software (version 7.6, Brazil). When the main effects were significant, differences among factor levels were tested for significance using the mean sum of square statement at P=0.05.

RESULTS

Growth Promoting Parameters

The results reveal that the microbial treatments not only enhanced the growth parameters in O. tenuiflorumbut also significantly modulated aromatic oil's quality. The physical parameters of soil indicate good soil health with a neutral pH (Table 1). Electrical conductivity is the ability of a material to transmit an electrical current and is expressed as milli-Siemens per meter (mS/m). In the present experiment, the pH and conductivity of the initial soil sample were 7.53 and 135 μScm^{-1}, respectively. According to USDA criteria, EC was lower than 400 μSm^{-1} indicating neutral stress of soil against the existing flora signifying for the good quality of the soil. The attributes like shoot height and weight, root length and weight, and so forth are not considered as an important character determining essential oil extraction but are momentous for growth promotion decision. The traits are also desirable as the flowering tops and young shoots of O. tenuiflorumare also used for herbal tea preparations. All of inoculation treatments increased shoot length which were significantly different ($p<0.05$) when compared with uninoculated control. All single and coinoculation treatments enhanced shoot length ranging from 88.33 cm to 95.67 cm (Figure 1). In the single inoculation, P. fluorescens (T3) treatment was observed with maximum effects as the plants depicted a 16.67 cm increase in shoot length over the control plants. The treatments had a lower variation coefficient of 1.52 (Table 2(a)) depicting the significant impact of all the treatments in enhanced growth. P. fluorescens intervention also resulted in the maximum increase in root length (13.8 cm) in alone treatments with an overall mean root length augmentation of 5.6 cm (9.7 to 15.3 cm). Fresh root weight was enhanced chiefly due to the combined treatment of P. fluorescens and B. megaterium (9.2 g). In the alone treatments, T. viride treatment depicted an augmentation of 2 cm over the control plants while the maximum shift of approximately 10 cm was observed in the combined treatment (T8) of all the bioinoculants, reflecting the growth promotional aspect of the organic treatments (Figure 1).

Table 1. Physical parameters of the soil

Physical parameters	Values
Electrical conductance (dsm. sup.$^{-1}$)	135 μS/cm
N	182.74 Kg/h
K	84 Kg/h
P	52 Kg/h
Na	205 Kg/h
Organic carbon	0.68%
pH	7.53

Table 2. (a) Relation between plant biomass and percentage essential oil in O. tenuiflorum cv. CIM-Ayu. (b) ANOVA table for biomass and percentage essential oil

(a)

Treatments	Shoot length (cm)	Root length (cm)	Root fresh weight (g)	Leaf yield (g)	Essential oil yield (%)
T1	71.00^{f}	9.67^{e}	5.33^{f}	160.83^{f}	0.56^{e}
T2	78.33^{e}	12.33^{d}	6.67^{e}	178.33^{e}	0.61^{d}
T3	87.67^{c}	13.83^{bc}	8.33^{c}	185.83^{d}	0.67^{c}
T4	78.00^{e}	13.17^{cd}	7.167^{de}	181.50^{e}	0.60^{d}
T5	81.67^{d}	13.33^{c}	7.67^{d}	187.17^{d}	0.67^{c}
T6	85.67^{c}	14.33^{b}	8.67^{b}	196.50^{c}	0.71^{b}
T7	91.67^{b}	14.50^{ab}	9.17^{ab}	203.50^{b}	0.77^{a}
T8	95.67^{a}	15.33^{a}	9.67^{a}	210.00^{a}	0.77^{a}
LSD ()	2.21	0.88	0.59	3.59	0.02
CV%	1.52	3.83	4.32	1.10	1.88

[a]Mean in each column followed by the same letters does not differ significantly ($p<0.05$) according to Duncan's multiple range test.

[b]Values followed by different small letters within columns are significantly different ($p<0.05$) according to the LSD test.

(b)

Source of variation	Degree of freedom (df)	Shoot length (cm)	Root length (cm)	Root fresh weight (g)	Leaf yield (g)	Essential oil yield (%)
Treatment	7	1358.96**	63.24**	42.50**	5026.79**	0.09**
Error	16	26.00	4.17	1.83	68.67	0.003
Total	23	1384.96	67.41	44.33	5095.46	0.09

**Significance at a level of 5% of probability (p<0.05).

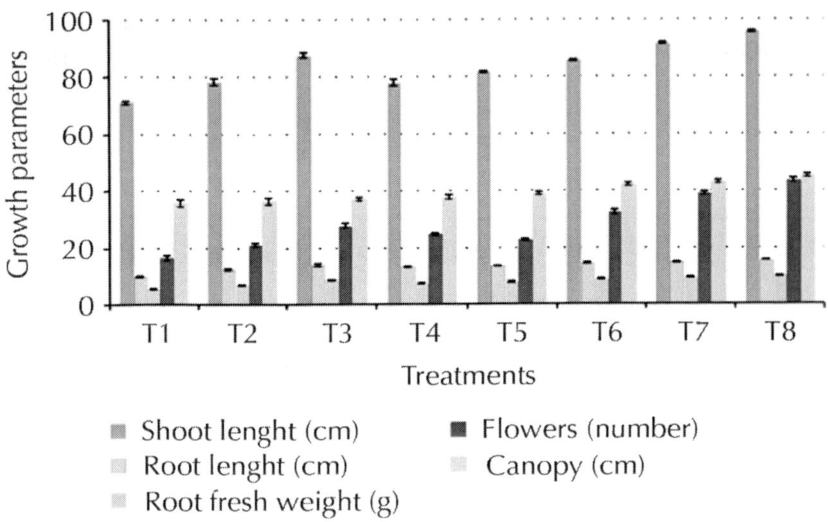

Figure 1. Growth parameters as influenced by the microbial treatments in Ocimum tenuiflorum cv. CIM-Ayu. a. Standard error of means (SEM) is indicated by the error bars.

Leaf Yield and Essential Oil Content (%w/v)

Leaf yield that is leaf fresh weight per plant is an important character in determining the aromatic oil yield. The microbial inoculants can vary the constituents of essential oil and thus determine its quality. Observations for leaf biomass depicted a maximum of 49.17 g increase

(T8) against the control (T1) plants. In the alone treatments, the highest leaf and oil yield were recorded in P. fluorescens treatment with an overall increase of 15.6% and 11%, respectively, against the control plants (Table 2(a)). Essential oil content showed a minimum of 21% enhancement in the combined treatments (T7 and T8) of microbial inoculants. Table 2(b) depicts a quick glance of ANOVA analysis for biomass and percentage essential oil with highly significant results as is indicated by the P values. Means were compared using the least significant difference (LSD) at probability level P=0.05 and revealed significant mean differences against the control plants. CV% values for leaf yield and percentage essential oil were 1.1 and 1.88, respectively, which showed greater variation among various treatments.

Identification of Essential Oil Components

To identify the mixture components, calculated retention indices (RI) were used. Retention indices of "neutral" components were compared with those reported (Adams 1995). When not less than three RI values are given for the same compound in different sources they were randomized. The results of Head Space analysis indicated that essential oil is rich in mono- and sesquiterpenoids. The major essential oil components as quantified through gas chromatography showed much variability. In alone treatments, the maximum increase (12.34%) in eugenol content was recorded in P. fluorescens treatment followed by B. megaterium treatment, showing 8.33% increase over the control plants (Figure 2). The combined treatment of P. fluorescens and T. viride enhanced eugenol content up to 56.22% as compared to 42.86% in control plants. Other essential oil constituents responsible for the characteristic aroma of O. tenuiflorum, namely, β -caryophyllene and methyl chavicol, were also appreciably enhanced by microbial treatments. A quick insight into major essential oil constituent for three best treatments is depicted through their chromatograms (Figure 3). The peaks for eugenol, methyl chavicol, and β-caryophyllene were identified at retention index (RI) of 26.6, 28.2, and 29.5 minutes, respectively.

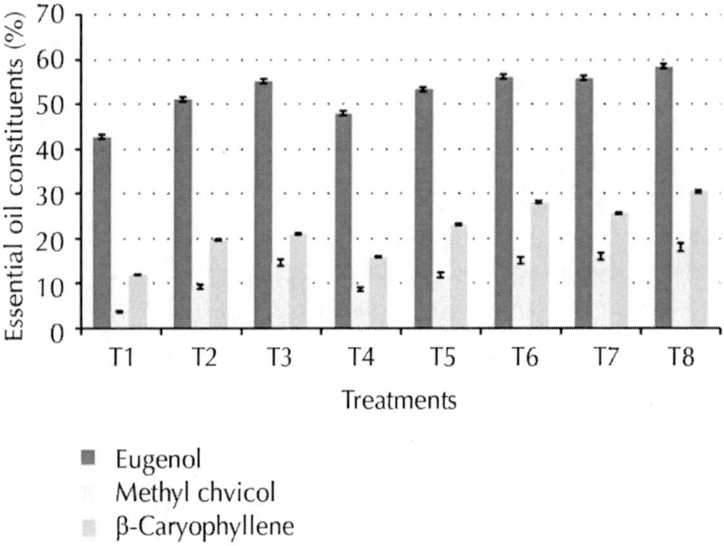

Figure 2. The major essential oil constituents influenced by different microbial treatments in Ocimum tenuiflorum cv. CIM-Ayu.

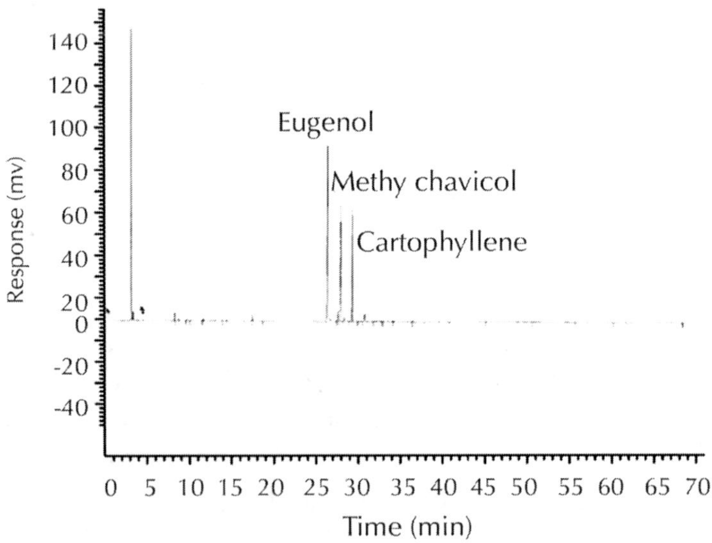

(a) Uninoculated control

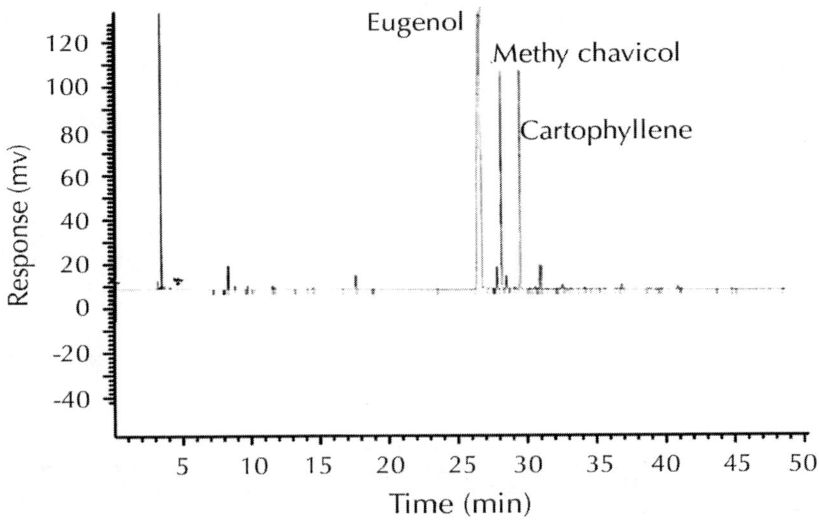

(b) Pseudomonas fluorescens

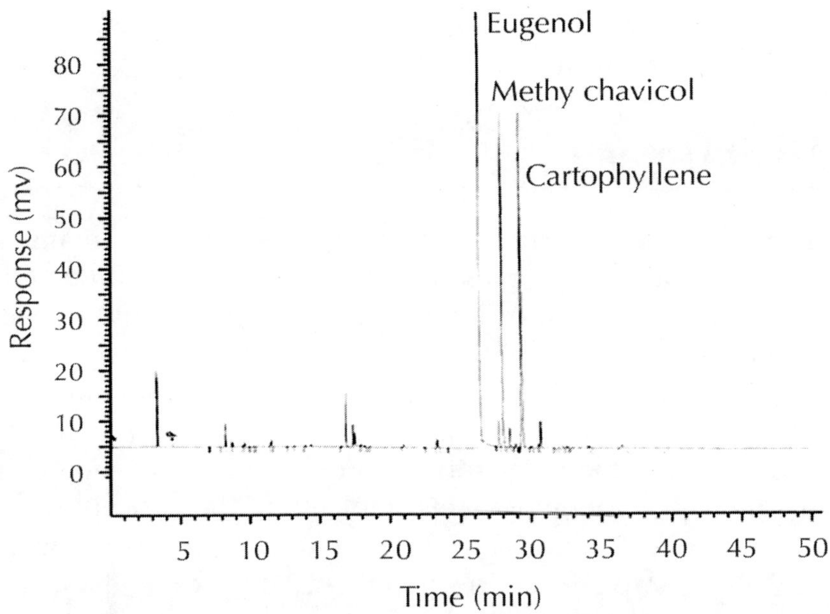

(c) Pseudomonas fluorescens + Trichderma viride

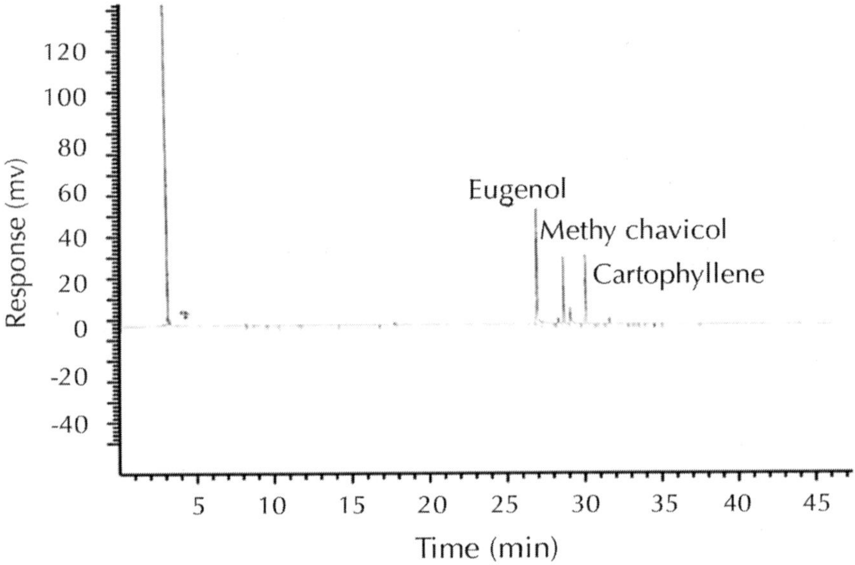

(d) Bacillus megaterium + Pseudomonas fluorescens +Trichoderma viride

Figure 3. GC chromatograms of selected microbial treatments depicting major essential oil constituents.

Nutrient Uptake

Nitrogen uptake was drastically enhanced in all the treatments with B. megaterium showing the maximum uptake of 2.83% followed by P. fluorescens as 2.62% in the alone treatments. The combined treatment of B. megaterium, P. fluorescens, and T. viride showed maximum uptake (2.93%) for nitrogen which was much higher than the control (1.67%) values. An enhanced uptake for nitrogen in all the treatments clearly shows that these microbial inoculants have an unambiguous role in growth promotion of plants (Table 3(a)). The phosphorus uptake results were quite contrasting to N_2 uptake as the phosphorus content declined in most of the alone treatments. However, the combined treatments showed a slight rise in phosphorus uptake with the highest phosphorus content recorded in the plants inoculated with the consortia of B. megaterium, P. fluorescens, and T. viride. An increased potassium uptake was also little influenced by the microbial treatments as is evident from the least significant difference and P values (Table 3(b)).

These inoculants also depicted an augmented uptake when combined with T. viride.

Table 3. (a) Effect of rhizospheric microbes on nutrient uptake in O. tenuiflorum cv. CIM-Ayu. (b) ANOVA analysis for nutrient uptake in O. tenuiflorum cv. CIM-Ayu

(a)

Treatments	Nutrient uptake (%)		
	N	P	K
T1	1.67[g]	0.32[d]	2.27[a]
T2	2.83[b]	0.28[ef]	2.30[a]
T3	2.62[c]	0.36[c]	1.62[d]
T4	2.23[d]	0.31[de]	1.93[c]
T5	2.13[e]	0.25[f]	1.85[c]
T6	2.07[f]	0.33[d]	2.08[b]
T7	2.11[e]	0.41[b]	2.24[a]
T8	2.93[a]	0.44[a]	2.33[a]
LSD (p=0.05)	0.04	0.03	0.11
CV%	0.98	5.50	2.96

[a]Mean in each column followed by same letters does not differ significantly ($p<0.05$) according to Duncan's multiple range test.

(b)

Source of variation	Degree of freedom (df)	N%	P%	K%
Treatment	7	3.99**	0.085**	1.37**
Error	16	0.01	0.01	0.06
Total	23	3.99	0.09	1.43

DISCUSSION

Medicinal plants constitute a large segment of the flora, which provide raw materials for pharmaceutical, cosmetic, and fragrance industries. Noteworthy increases in growth and yield of vital crops can be the outcome of plant growth promoting rhizobacteria (PGPR) inoculation. PGPRs actively colonize plant roots and increase plant growth and yield [19]. In the present study, growth parameters were significantly enhanced by the microbial treatments with an improved oil quality and quantity.

Glick et al. [20] considered plant growth promoting rhizobacteria as a better alternative to the established chemical strategies for facilitating plant growth. Rhizospheric microbes influence plant growth through the release of several antibiotics such as celastramycins A-B [21], kakadumycins [22], and demethylnovobiocins [23]. Furthermore PGPRs are also known to secrete a number of bioactive metabolites, for example, Taxol, maytansinoids. PGPRs facilitate the production of plant growth promotion compounds, such as auxins, cytokinins, and gibberellins, or by producing siderophore to bind Fe^{3+} from the environment and help to improve nutrient uptake, supply of plant nutrients (nitrogen, phosphate, and other mineral nutrients), or suppression of stress ethylene production by 1-aminocyclopropane-1-carboxylate (ACC) deaminase activity [24].

The present experimentation evaluates the effect of plant growth promoting rhizobacteria (PGPR) on nutrient uptake, growth, and yield attributes in Basil (O. tenuiflorum). The estimated impacts of microbial treatments on plant growth indices, mineral content, and essential oils in combined treatments validate the consortia inoculation to be a better option for enhancing growth parameters and yield attributes. Nitrogen and phosphorus are the two major plant nutrients responsible for influencing vegetative and reproductive phase of the plant growth, respectively. The colonization of root by PGPRs stimulates increased nutrient uptake at the root interface which results in better absorption of water and nutrients from the soil [25]. In the present study, nitrogen content in the biomass of the inoculated plants was much higher and it was significantly high in B. megaterium treatments. Bacillus spp. being a free living N_2-fixer supplements more nitrogen to biomass during the growth. In addition to the improvement of plant growth, these microbes also improve the soil fertility by aggregation and adding nutrients to soil. Similarly better response of O. basilicum plants inoculated with consortia of Glomus fasciculatum, P. fluorescens, and B. megaterium has been reported [26]. Thus it is suggested that microbial treatment improves the plant growth, biomass, and yield by supplementing plant nutrients and producing growth hormones.

Increased nutrient uptake by plants inoculated with PGPRs has been attributed to the production of plant growth regulators at the root interface, which stimulated root development and resulted in better absorption of water and nutrients from the soil as PGPRs are able to solubilise "unavailable" forms of minerals by excreting organic acids which dissolve or chelate mineral ions to soluble form, thereby improving the competitiveness and external stress responses of the plant. The results in our experiment indicated a higher response with all the microbial treatments especially in combined microbial treatments. Essential oils are commonly extracted from aromatic crops through steam distillation. Since the method requires huge amounts of biomass, the microbial treatment proves to be beneficial for improved biomass yield. The GC-analysis of the oil having a combined treatment of B. megaterium, P. fluorescens, and T. viride revealed a major compound (58.5%) with a Kovat's index of 1356. The compound was identified as eugenol, a monoterpene having antimicrobial, insecticidal, antihelminthic, and nematicidal properties. Methyl chavicol also possesses antifungal and antibacterial activities.

CONCLUSIONS

The results reveal that microbial association not only enhanced the growth parameters in O. tenuiflorum but also significantly modulated aromatic oil's quality. The observations explained appreciable correlations between functional rhizospheric microbes and growth promoting traits, namely, plant height, shoot and root fresh weight, N, P, and K uptake, leaf yield and essential oil content, and major oil constituents. The study also emphasises that P. fluorescens greatly alters the growth parameters in O. tenuiflorum. With this experiment as primary step, further study is needed to develop biofertilizers consortium for commercially grown medicinal plants.

ACKNOWLEDGMENTS

The authors wish to thank the Director, CSIR-CIMAP, Lucknow (India), for providing necessary facilities and Dr. D. Saikia for his editorial suggestions.

REFERENCES

1. A. A. Farooqi and B. S. Sreeramu, Cultivation of Medicinal and Aromatic Crops, University Press, Hyderabad, India, 2001.
2. R. K. Lal, S. P. S. Khanuja, A. K. Agnihotri et al., "High essential oil and eugenol yielding cultivar ofOcimum sanctum "CIM-AYU"," US Patent 2005/0091705 A1, 2005.
3. F. A. Atta and O. A. O. Saad, "Biofertilizers as potential alternative of chemical fertilizer forCathranthus roseus," Journal of Agricultural Science, vol. 26, pp. 7193–7208, 2001.
4. M. M. Wagenaar and J. Clardy, "Dicerandrols, new antibiotic and cytotoxic dimers produced by the fungus Phomopsis longicolla isolated from an endangered mint," Journal of Natural Products, vol. 64, no. 8, pp. 1006–1009, 2001.
5. S. M. Nadeem, I. Hussain, M. Naveed, H. N. Ashgar, Z. A. Zahir, and M. Arshad, "Performance of plant growth promoting

rhizobacteria containing ACC-deaminase activity for improving growth of maize under salt-stressed conditions," Pakistan Journal of Agricultural Sciences, vol. 43, pp. 114–121, 2006.

6. J. W. Kloepper, R. Lifshitz, and R. M. Zablotowicz, "Free-living bacterial inocula for enhancing crop productivity," Trends in Biotechnology, vol. 7, no. 2, pp. 39–44, 1989.

7. B. Giri, P. H. Giang, R. Kumari, R. Prasad, and A. Varma, "Microbial diversity in soils," in Roles in Genesis and Functions, F. Buscot and S. Varma, Eds., pp. 195–212, Springer, Heidelberg, Germany, 2005.

8. J. Barriuso, B. Ramos Solano, C. Santamaría, A. Daza, and F. J. Gutiérrez Mañero, "Effect of inoculation with putative plant growth-promoting rhizobacteria isolated from Pinus spp. on Pinus pinea growth, mycorrhization and rhizosphere microbial communities," Journal of Applied Microbiology, vol. 105, no. 5, pp. 1298–1309, 2008.

9. J. Vacheron, G. Desbrosses, M. L. Bouffaud et al., "Plant growth promoting rhizobacteria and root system functioning," Frontiers in Plant Science, vol. 4, p. 356, 2013.

10. J. W. Kloepper, C. Ryu, and S. Zhang, "Induced systemic resistance and promotion of plant growth byBacillus spp.," Phytopathology, vol. 94, no. 11, pp. 1259–1266, 2004.

11. L. Pessoa, M. McKenna, E. Gutierrez, and L. G. Ungerleider, "Neural processing of emotional faces requires attention," Proceedings of the National Academy of Sciences of the United States of America, vol. 99, no. 17, pp. 11458–11463, 2002.

12. R. Harisaranraj, S. S. Babu, and K. Suresh, "Callus induction and plant regeneration of Vigna mungo(L.) Hepper via half seed explant," Ethnobotanical Leaflets, vol. 12, pp. 577–585, 2008.

13. B. V. Subbiah and G. L. Asija, "A rapid procedure for the estimation of available nitrogen in soils,"Current Science, vol. 25, pp. 259–260, 1956.

14. R. Pandey, A. Kalra, M. L. Gupta, and P. Sharma, "Phytonematodes: Major pest of MAPs," inProceedings of the 1st National Interactive Meet on Medicinal and Aromatic Plants, S. Mathur, Ed., pp. 188–197, CIMAP, Lucknow, India, 2003.

15. S. K. Saikia, S. Tiwari, and R. Pandey, "Rhizospheric innovations for growth enhancement andMeloidogyne incognita management in Mentha arvensis cv. Kosi," International Journal of Environmental Science and Technology, vol. 3, no. 1, pp. 26–34, 2012.
16. R. Pandey, A. Gupta, H. N. Singh, and A. Kalra, "Phytonematode management through Bacteria: an Underground Battle for existence," in Recent Advances. Biopesticides Biotechnological Applications, J. K. Johri and N. B. R. I. Lucknow, Eds., pp. 1–26, 2009.
17. M. L. Jackson, Soil Chemical Analysis, Prentice Hall of India, New Delhi, India, 1973.
18. G. W. Snedecor and W. G. Cochran, Statistical Methods, Iowa State University Press, 8th edition, 1989.
19. S. Shawky, R. Z. El-shennawy, and A. M. Shady, "Biological control of Meloidogyne javaniica on tomato plants with isolated bioagent in Egypt," The Journal of Agricultural Science, Mansoura University, vol. 37, pp. 6049–6063, 2006.
20. B. R. Glick, B. Todorovic, J. Czarny, Z. Cheng, J. Duan, and B. McConkey, "Promotion of plant growth by bacterial ACC deaminase," Critical Reviews in Plant Sciences, vol. 26, no. 5-6, pp. 227–242, 2007.
21. C. Pullen, P. Schmitz, K. Meurer et al., "New and bioactive compounds from Streptomyces strains residing in the wood of Celastraceae," Planta, vol. 216, no. 1, pp. 162–167, 2002.
22. U. Castillo, J. K. Harper, G. A. Strobel et al., "Kakadumycins, novel antibiotics from Streptomyces sp. NRRL 30566, an endophyte of Grevillea pteridifolia," FEMS Microbiology Letters, vol. 224, no. 2, pp. 183–190, 2003.
23. Y. Igarashi, "Screening of novel bioactive compounds from plant-associated actinomycetes,"Actinomycetologica, vol. 18, pp. 63–66, 2004.
24. V. Kannan and R. Sureendar, "Synergistic effect of beneficial rhizosphere microflora in biocontrol and plant growth promotion," Journal of Basic Microbiology, vol. 49, no. 2, pp. 158–164, 2009.
25. P. Nimnoi, N. Pongsilp, and S. Lumyong, "Endophytic actinomycetes isolated from Aquilaria crassnaPierre ex Lec and

screening of plant growth promoters production," World Journal of Microbiology and Biotechnology, vol. 26, no. 2, pp. 193–203, 2010.

26. B. A. Iwalokun, G. O. Gbenle, T. A. Adewole, S. I. Smith, K. A. Akinsinde, and E. O. Omonigbehin, "Effects of Ocimum gratissimum L. essential oil at subinhibitory concentrations on virulent and multidrug-resistant Shigella strains from Lagos, Nigeria," Acta Pathologica, Microbiologica et Immunologica Scandinavica, vol. 111, no. 4, pp. 477–482, 2003.

Chapter 7

Dual Role of Hydrogen Peroxide in Arabidopsis Guard Cells in Response to Sulfur Dioxide

Huilan Yi,[1] Xin Liu,[1,2] Min Yi,[1,3] and Gang Chen[2]

[1]School of Life Science, Shanxi University, Taiyuan 030006, China

[2]Shanxi Provincial Guoxin Energy Development Group Co., LTD., Taiyuan 030006, China

[3]Department of Statistics, University of Missouri-Columbia, Columbia, MO 65211, USA

ABSTRACT

Sulfur dioxide (SO_2) is a major air pollutant and has significant impacts on plant physiology. Plant can adapt to SO_2 stress by controlling stomatal movement, gene expression, and metabolic changes. Here we show clear evidences that SO_2-triggered hydrogen peroxide (H_2O_2) production mediated stomatal closure and cell death in Arabidopsis leaves. High levels of SO_2 caused irreversible stomatal closure and decline in guard cell viability, but low levels of SO_2 caused reversible stomatal closure. Exogenous antioxidants ascorbic acid (AsA) and catalase (CAT) or Ca^{2+} antagonists EGTA and $LaCl_3$ blocked SO_2-induced stomatal closure and decline in viability. AsA and CAT also blocked SO_2-induced H_2O_2 and $[Ca^{2+}]_{cyt}$ elevation. However, EGTA and $LaCl_3$ inhibited SO_2-induced $[Ca^{2+}]_{cyt}$ increase but did not suppress SO_2-induced H_2O_2 elevation. These results indicate that H_2O_2 elevation triggered stomatal closure and cell death via $[Ca^{2+}]_{cyt}$ signaling in SO_2-stimulated Arabidopsis guard cells. NADPH oxidase inhibitor DPI blocked SO_2-induced cell death but not the stomatal closure triggered by low levels of SO_2, indicating that NADPH oxidase-dependent H_2O_2 production plays critical role in SO_2 toxicity but is not necessary for SO_2-induced stomatal closure. Our results suggest that H_2O_2 production and accumulation in SO_2-stimulated plants trigger plant adaptation and toxicity via reactive oxygen species mediating Ca^{2+} signaling.

INTRODUCTION

Sulfur dioxide (SO_2) is a harmful gas that is emitted largely from burning coal, high-sulfur oil, and fuels. During the past few decades, the concentration of SO_2 in the atmosphere has increased in many areas of the world, especially in the developing countries. High levels of SO_2 can injure many plant species and varieties, resulting in photosynthesis decline, growth inhibition, and even death [1–4]. Sulfur dioxide enters plants mainly through the open stomata [5]. Once it enters the leaf, SO_2 is hydrated to form HSO_3^- and SO_3^{2-}. The toxicity of SO_2 is derived from molecular species sulfite (SO_3^{2-}) and bisulfite (HSO_3^-) generated after SO_2 is dissolved in cellular fluid [6]. Sulfite oxidation, which is the detoxification reaction of sulfite to sulfate (SO_3^{2-}), leads to the formation of reactive oxygen species (ROS) in plant cells [7, 8]. The production and accumulation of ROS are one of the key events in plant response to SO_2 [9–12].

ROS have been proposed as central components of plant response to both biotic and abiotic stresses. Under such conditions, ROS may play two very different roles: exacerbating damage or signaling the activation of defense responses [13, 14]. It has been reported that ROS act as signaling molecules mediating a variety of physiological responses, including stomatal movement and gene expression [15–17], although they can attack biomolecules such as nucleic acids, proteins, and lipids leading to cell damage and death [16, 18, 19]. Results of previous studies have shown that plants can adapt to SO_2 stress by controlling stomatal movement and gene transcription [11, 12, 20]. Stomatal closure could protect the leaves against further entry of the environmental SO_2, while differential gene expression could regulate the metabolic routes of plant cells providing long-term adaptation to environmental stress. However, up to now, it is not clear if the ROS production is closely associated with the initial physiological mechanisms responsible for plant responses to SO_2 stress.

ROS overproduction can trigger plant adaptation or cell damage during abiotic stress. However, there has been little overlap between these two fields of research. In the present study, guard cells of A. thaliana leaves, which are a well-developed model cell system for studying the signal transduction in plant cells [21–23], were employed to investigate the cellular mechanism of plant response to SO_2 stress. To the best of our knowledge, this is the first report of H_2O_2 mediating both adaptation response (stomatal movement) and cytotoxicity (cell death) in plant response to SO_2 stress. Our results show that H_2O_2 elevation triggered both stomatal closure and cell viability loss via Ca^{2+} signaling in SO_2-treated plants.

MATERIALS AND METHODS

Plant Material Preparation

Plants of Arabidopsis thaliana (L.) ecotype Columbia (Col-0) were grown in soil (Klasmann-Deilmann) in temperature-controlled growth rooms at 22°C with an average light intensity of 240 $\mu mol L m^{-1} s^{-1}$, a 16 h photoperiod per day, and 60% relative humidity.

Young fully expanded leaves were harvested from 4-week-old Arabidopsis plants. The abaxial epidermes were peeled from the underside of each leaf and cut into small pieces. The isolated epidermal strips were immediately floated in 10 mM 2-(N-morpholino)ethanesulfonic acid (MES; Bio Basic Inc.) buffer (10 mM MES-Tris, pH 7.0, and 50 mM KCl).

Determination of Stomatal Aperture

The isolated epidermal strips were incubated in 10 mM MES buffer for 2 h, for stomata-opening under continuous light of 180 μmol m^{-2} s^{-1} at 22°C or for stomata-closuring in the dark, and then incubated for 2 h in MES buffer containing a certain amount of SO_2 hydrates (a mixture of sodium sulfite and sodium bisulfite, 3 : 1 mM/mM, prepared freshly before use) under continuous illumination of 180 μmol m^{-2} s^{-1} at 22°C. Control samples were treated under the same conditions with 10 mM MES buffer. The isolated epidermal strips were treated with a mixture of SO_2 hydrates and a certain amount of antagonists which include NADPH oxidase inhibitor diphenylene iodonium (DPI, Sigma), antioxidants catalase (CAT, Sigma), and ascorbic acid (AsA, Sigma) and Ca^{2+} antagonists $LaCl_3$ and ethylene glycol tetraacetic acid (EGTA, Sigma) to examine the protective effects. After 2 h of chemical exposure, the epidermal strips were mounted on a microscopy slide, moistened with 10 mM MES buffer, and covered with a slip. Stomatal aperture was measured by using a digital microscope camera system (DP72, Olympus) and an attached DP2-BSW software. At least three leaves and 300 stomata per leaf were measured in each treatment and all of the experiments were independently repeated at least three times. For the recovery groups, after 2 h of chemical exposure, the isolated strips were resuspended in MES buffer for 2 h under continuous light followed by stomatal aperture measurement. For time-course experiment, the stomatal apertures were examined every five minutes after isolated strips were incubated in SO_2 hydrates.

Determination of Cell Viability

Cell viability was assessed by using the method of double staining with fluorescein diacetate (FDA; Bio Basic Inc.) and propidium iodide

(PI; Sigma). After 2 h of chemical exposure, the epidermal strips were simultaneously stained with 0.1 mg·L^{-1} FDA and 10 μM PI for visualizing cell viability. At least three leaves and 300 guard cells per leaf were observed in each treatment and all of the experiments were independently repeated three times.

Measurement of Reactive Oxygen Species

Reactive oxygen species in guard cells of epidermal peels were detected using 2′,7′-dichlorodihydrofluorescein diacetate (DCFH-DA; Beyotime) according to the methods described by Yi et al. [24]. After 2 h of exposure to the chemicals, the epidermal strips were incubated in 20 μM DCFH-DA for 30 min in the dark. The mean value of fluorescence intensity, resulting from 600 guard cells in three independent experiments, represents the intracellular H_2O_2 level for each treatment.

Measurement of Intracellular Ca^{2+}

The concentration of intracellular Ca^{2+} ($[Ca^{2+}]_{cyt}$) was measured using fluo-3 acetomethoxyester (Fluo-3 AM; Beyotime) as described in our previous report [24]. The average fluorescence intensity of Fluo-3 AM obtained from three different leaves and 200 guard cells represents the $[Ca^{2+}]_{cyt}$ level for each treatment. All experiments were independently repeated three times.

Statistical Analysis

All values of mean and standard deviation (SD) were obtained from three independent experiments. Analysis of variance (ANOVA) and Dunnett's t -test were used to determine the significant differences among the control and a series of treatment groups.

RESULTS

SO_2-Evoked H_2O_2 Production Is Involved in the Regulation of Stomatal Movement

As shown in Figure 1, SO_2 hydrates promote stomatal closure and inhibit light-promoted stomatal opening in Arabidopsis leaves. The width of stomatal aperture decreased in a concentration-dependent manner and showed a significant decrease after the isolated strips were exposed for 2 h to SO_2 hydrates at concentrations of 10 to 500 μM. However, stomatal closure evoked by low SO_2 concentrations (below 200 μM) could be reversed completely after SO_2 removal; otherwise the decline of stomatal aperture in high SO_2 concentration (500 μM) group could be partly reversed.

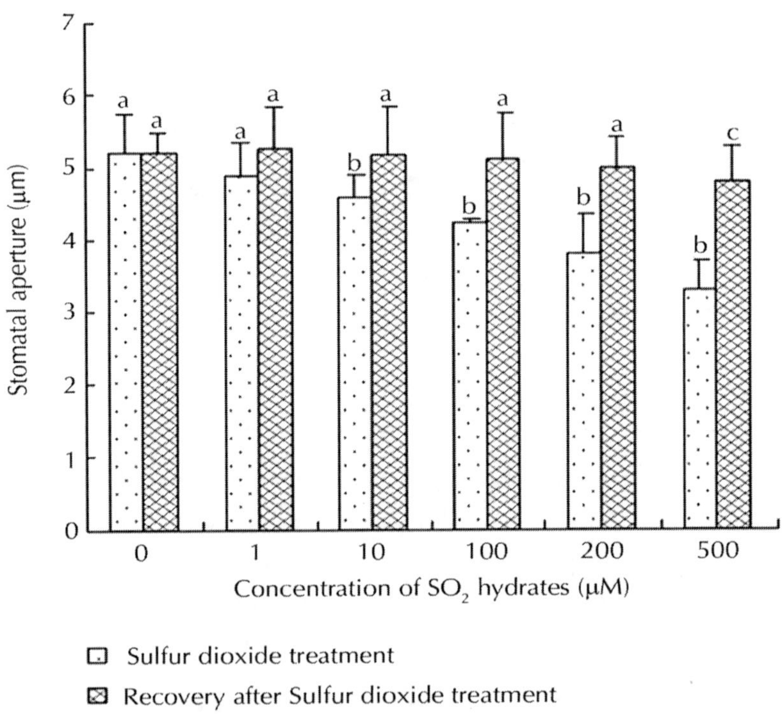

(a)

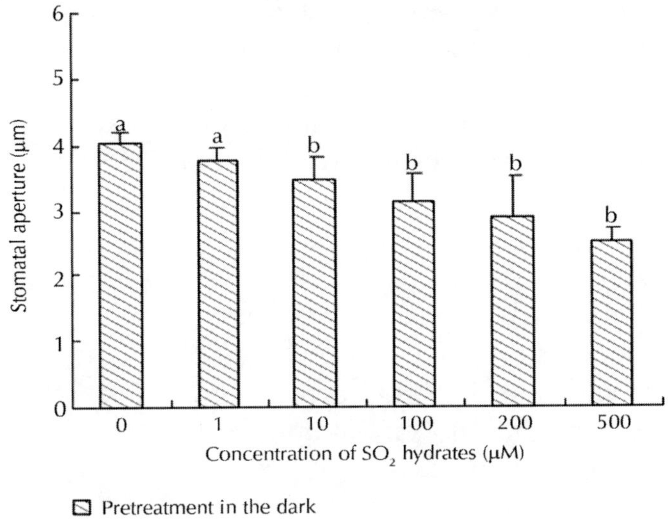

(b)

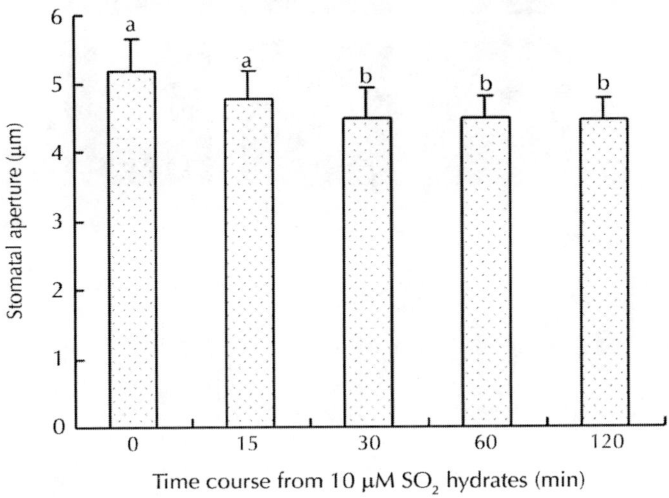

(c)

Figure 1. Promotion of stomatal closure by SO_2 and its reversibility upon SO_2 removal. Stomata of Arabidopsis leaf epidermis were allowed to open in light [(a) and (c)] or to close in the dark (b) for 2 h and then were incubated for 2 h in SO_2 hydrates. b, A significant difference at from control. c, A significant difference at $P<0.05$ between SO_2treatment and its recovery group.

Time course analysis of stomatal movement showed that the diameter of the stomatal aperture decreased markedly within the first 30 minutes of exposure to 10 μM SO_2 hydrates and continued to decline at a slower rate in the remaining 90 minutes of SO_2 exposure (Figure1(c)). SO_2-induced stomatal closure was associated with an elevated H_2O_2 level in Arabidopsis guard cells. Exposure to SO_2 hydrates not only caused smaller stomatal apertures (Figure 1) but also evoked increased H_2O_2 level (1.2- to 1.7-fold) in Arabidopsis guard cells (Figure 2). When H_2O_2 elevation was blocked by exogenous H_2O_2 scavengers CAT or AsA, SO_2-induced stomatal closure was efficiently reversed (Figure 2). These results indicate that H_2O_2production in SO_2-stimulated guard cells mediates stomatal movement upon SO_2 stress.

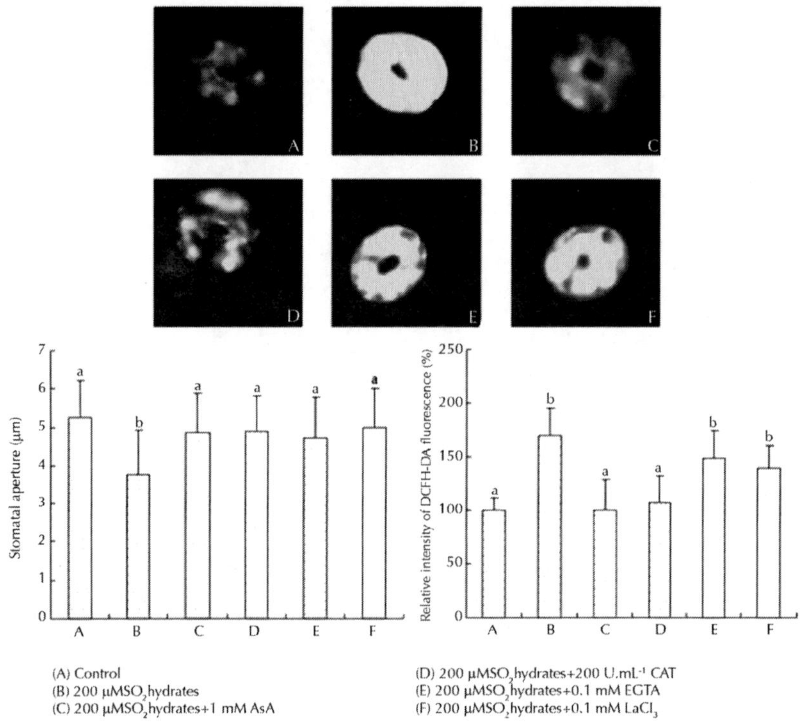

Figure 2. Effects of antioxidants and calcium antagonists on SO_2-induced stomatal closure and H_2O_2 elevation in Arabidopsis guard cells. Different superscript letters indicate significant differences (P<0.01). The same letters indicate no significant difference. The green fluorescence (DCFH-DA) of guard cells indicates H_2O_2 content.

SO_2-Evoked H_2O_2 Elevation Triggered Cell Death

To investigate the role of H_2O_2 in irreversible stomatal closure in response to SO_2, we detect the viability of Arabidopsis guard cells by double staining with fluorescein diacetate (FDA) and propidium iodide (PI). The results showed that SO_2 induced guard cell death in a concentration-dependent manner (Figure 3). Cell death reached 24% in 6 mM SO_2 hydrates treatment group, but no obvious cell death could be observed in 10 to 200 μM SO_2 hydrates treatment groups. The H_2O_2 level of guard cells exposed to 2 to 6 mM SO_2 hydrates showed a statistically significant increase (1.90- to 2.30-fold), which is higher than those exposed to 10 to 500 μM SO_2 hydrates. Moreover, exposure to SO_2 hydrates simultaneously with 200 U·mL^{-1}CAT or 0.1 mM AsA, SO_2-induced cell death was efficiently blocked, associated with a significant decrease in H_2O_2 level of guard cells (Figure 3). These results clearly demonstrate that an elevated H_2O_2 level can trigger guard cell death leading to stomatal dysfunction and irreversible stomatal closure in SO_2-treated Arabidopsis leaves.

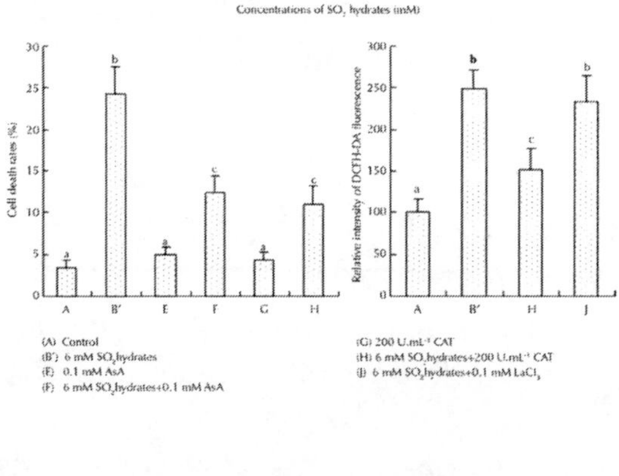

Figure 3. Sulfur dioxide caused viability decline associated with H_2O_2 elevation in Arabidopsis guard cells. Different superscript letters indicate significant differences ($P<0.01$). The same letters indicate no significant difference.

To further confirm the role of H_2O_2 in SO_2 toxicity, we investigate the protective effects of CAT on cell viability in SO_2-treated samples. The results showed that application of 1000 U·mL^{-1} CAT could completely block the cell death evoked by 2 mM and lower concentrations of SO_2 hydrates, but the cytotoxicity evoked by 6 mM SO_2 hydrates was only partly blocked (Figure 4). These results clearly demonstrate that H_2O_2 production, which may work together with other molecules, is enough to trigger SO_2 toxicity.

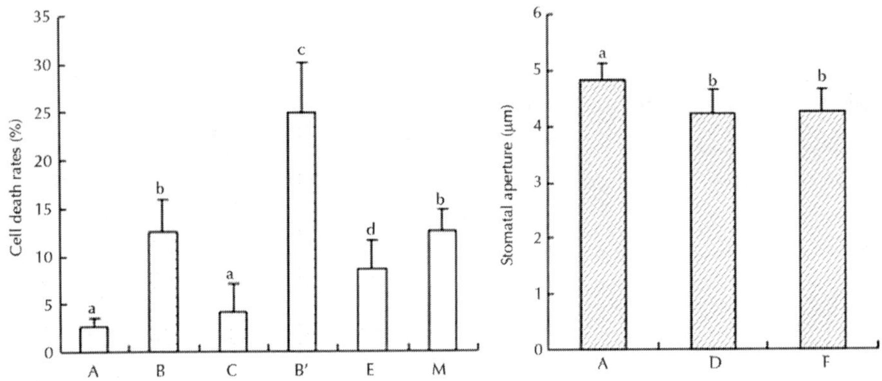

(A) Control
(B) 2mM SO_2 hydrates
(C) 2mM SO_2 hydrates + 1000 U mL^{-1} CAT
(B′) 6mM SO_2 hydrates
(E) 6mM SO_2 hydrates + 1000 U.mL^{-1} CAT
(M) 6 mM SO_2 hydrates+20 20 μM DPI
(D) 10 μM SO_2 hydrates
(F) 10 μM SO_2 hydrates + 20 μM DPI

Figure 4. Hydrogen peroxide scavenger and NADPH oxidase inhibitor DPI blocked SO_2-induced cell death, but DPI cannot block SO_2-induced stomatal closure. Different superscript letters indicate significant difference at $P<0.01$. The same letters indicate no significant difference.

H_2O_2 Action Is Dependent on Its Concentrations and Spatial Generation Patterns

As shown above, SO_2 can cause stomatal closing and decline in cell viability, which is dependent on SO_2 concentrations and H_2O_2 level. However, there was no clear dividing line between a safe level and a toxic level. In order to understand what constitutes a safe level, an inhibitor of NADPH oxidase was used to detect its inhibitory effects. The results show that application of 20 μM NADPH oxidase inhibitor

DPI for 2 h markedly blocked cell death evoked by 2 mM SO_2 hydrates. But 20 μM DPI did not inhibit stomatal closing evoked by 10 μM SO_2 hydrates (Figure 4). These findings indicate that SO_2-triggered stomatal closing can be driven by NADPH oxidase-dependent and -independent H_2O_2 generation, but NADPH oxidase-dependent H_2O_2 generation is involved in SO_2-caused cytotoxicity.

H_2O_2 Acts Upstream of Ca^{2+} Signaling in Response to SO_2 Stress

SO_2 exposure enhanced the fluorescence intensity of Fluo-3 AM (Ca^{2+} indicator) inArabidopsis guard cells. The relative fluorescence intensity of Fluo-3 AM, resulting from more than 600 guard cells per treatment group in three independent experiments, increased obviously in guard cells exposed to 10 μM to 6 mM SO_2 hydrates for 2 h but decreased markedly when isolated strips were treated simultaneously with SO_2 hydrates and 0.1 mM calcium channel blocker $LaCl_3$ (Figure 5). These results indicate that SO_2 exposure evokes an elevation of $[Ca^{2+}]_{cyt}$ level, and Ca^{2+} influx through Ca^{2+} channels in the plasma membrane results in $[Ca^{2+}]_{cyt}$ elevation.

As shown in Figures 2 and 5, the addition of $LaCl_3$ and calcium chelator EGTA to SO_2hydrates blocked $[Ca^{2+}]_{cyt}$ elevation, stomatal closure, and cell death evoked by SO_2. Stomatal closure and cell death occurred with increased $[Ca^{2+}]_{cyt}$ in SO_2-treated samples, but both of them were significantly suppressed by 0.1 mM $LaCl_3$ or EGTA. These results indicate that elevated $[Ca^{2+}]_{cyt}$ level is an important stimulus driving stomatal movement and cytotoxicity.

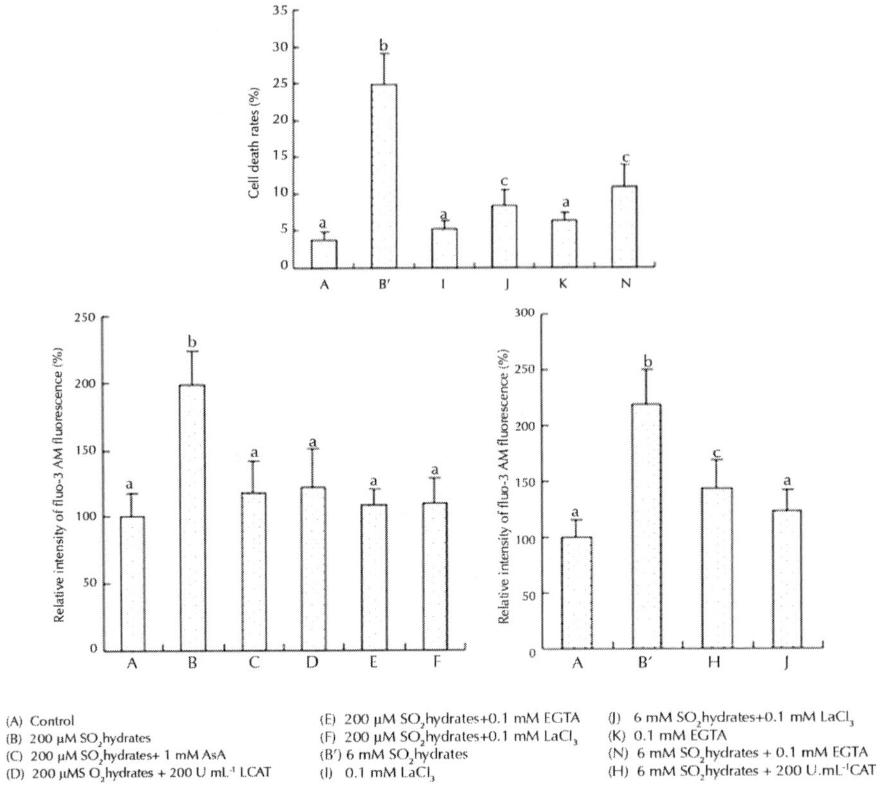

Figure 5. Effects of antioxidants and Ca^{2+} antagonists on SO_2-induced Ca^{2+} elevation and cell death in Arabidopsis guard cells. Different superscript letters indicate significant differences (P<0.01). The same letters indicate no significant difference.

To understand the pathways to plant responses, we study the signaling pathways in Arabidopsis guard cells. As shown in Figures 2, 3, and 5, both antioxidants AsA and CAT and Ca^{2+} antagonists $LaCl_3$ and EGTA can suppress SO_2-induced stomatal closure and cell death, but application of H_2O_2 scavenger CAT (200 U·mL^{-1}) decreased $[Ca^{2+}]_{cyt}$ elevation evoked by SO_2, whereas application of Ca^{2+} channel inhibitor $LaCl_3$ (0.1 mM) did not affect H_2O_2elevation evoked by SO_2. These results confirm the involvement of Ca^{2+} downstream of H_2O_2production, indicating that H_2O_2 triggers stomatal movement and cell death via Ca^{2+}signaling in plant response to SO_2 stress.

DISCUSSION

Plant could adapt to environmental challenges through various means. Our recent findings showed that SO_2 fumigation caused an increased ROS production accompanied with differential gene expression and stomatal closure in Arabidopsis plants [12, 20]. The results of the present study show the evidences that H_2O_2 acts as an important signaling molecule triggering stomatal movement and cell death in plant response to SO_2. SO_2-caused cell viability decline could interfere with the normal function of stomata to further affect plant physiology under environmental stress. However, the occurrence of SO_2-caused guard cell death was associated with decreased stomatal aperture, suggesting the existence of a complex signaling network in plant responses to environmental stress.

Environmental challenges including biotic and abiotic stresses could induce ROS production in plant cells [25]. It has been documented that ROS play an important role in signal transduction of stomatal movement regulation and gene expression activation in plant response to environmental stresses [26–29]. ROS, which can be used as rapid long-distance autopropagating signals that are transferred throughout the plant in response to different environmental conditions, are widely considered to be an important player in guard cell signaling [26, 30]. The data presented above indicate a requirement for H_2O_2 production in plant response to SO_2 stress. First, SO_2 induces stomatal closure and cell death associated with an increased intracellular H_2O_2 level (Figures 2 and 3). Second, two types of H_2O_2 scavengers CAT and AsA could block SO_2-evoked stomatal closure and cell death; in particular, the effects of SO_2 at low concentrations could be completely reversed by CAT (Figures 2 and 4). Third, H_2O_2 scavenger blocks the SO_2-induced increase in intracellular Ca^{2+} required for stomatal closure and toxicity (Figure 5). Our study, validating the strong positive correlation between intracellular H_2O_2 level and stomatal closure/cell death, demonstrates a key role of H_2O_2 as trigger of stomatal closure and/or guard cell death in response to SO_2.

Sulfur dioxide inhibited light-promoted stomatal opening and promoted stomatal closure leading to the smaller size of Arabidopsis stomata. High concentrations of SO_2 also caused viability loss of Arabidopsis guard cells. The time course experiments show that H_2O_2 level of guard cells increased gradually during SO_2 exposure (date not shown); therefore, an increased H_2O_2 level in SO_2-stimulated Arabidopsis cells might trigger stomatal closure firstly and then cell viability loss through ROS-mediated cell death pathway as shown in V. faba cells [24] or through the gradual accumulation of free radical damage to biomolecules during SO_2exposure.

There are two primary sources of H_2O_2 in guard cells: chloroplasts and plasma membrane-associated NADPH oxidase [22, 31, 32]. The results of the time course experiments with single cell assays using the fluorescent probe DCFH-DA showed that H_2O_2 generation was dependent on SO_2 concentration and that the increase in fluorescence intensity of chloroplasts occurred significantly earlier than within the other regions of guard cells (date not shown), demonstrating an enhanced H_2O_2 production in chloroplasts of guard cells. NADPH oxidase in plasma membrane contributes to $O^{2\cdot}$ generation and ROS elevation in plant responses to abiotic stresses [33]. Application of DPI, which is widely accepted as a relatively specific direct inhibitor of NADPH oxidase, markedly blocked SO_2-induced cell death but cannot suppress stomatal closure evoked by low concentrations of SO_2. These findings indicate that H_2O_2-mediated stomatal closure in Arabidopsis leaves exposed to low concentrations of SO_2 is not dependent on the activity of plasma membrane NADPH oxidase, but cell death evoked by high concentrations of SO_2 is NADPH oxidase-dependent.

SO_2 exposure caused stomatal closure and guard cell death associated with $[Ca^{2+}]_{cyt}$ elevation, whereas application of either Ca^{2+} chelator EGTA or Ca^{2+} channel inhibitor $LaCl_3$ blocking SO_2-evoked $[Ca^{2+}]_{cyt}$ elevation, stomatal closure and cell death evoked by SO_2 were effectively blocked. These results demonstrate that a channel-mediated Ca^{2+} influx across the plasma membrane contributes to the elevation of $[Ca^{2+}]_{cyt}$ and subsequent stomatal closure and cell death in SO_2-stimulated guard cells. However, it is not clear how Ca^{2+} signaling recognizes the different situations of cells to mediate appropriate processes such as stomatal closure and cell death.

It has been found that H_2O_2 could activate plasma membrane Ca^{2+} channels leading to $[Ca^{2+}]_{cyt}$ increase in plant cells [34–36]. Therefore, H_2O_2 activation of plasma membrane Ca^{2+} channels may be a central step in SO_2-induced stomatal closure and/or cell death. The results of our present study also showed that application of antioxidant CAT and AsA significantly decreased SO_2-evoked $[Ca^{2+}]_{cyt}$ elevation, but application of Ca^{2+} channel blocker $LaCl_3$ did not affect SO_2-evoked H_2O_2 increase, indicating that H_2O_2 acts upstream of Ca^{2+} signaling in SO_2-induced stomatal closure and/or cell death. These observations were consistent with other previous reports that a $[Ca^{2+}]_{cyt}$ increase was linked to H_2O_2 production and was involved in ROS-mediated stomatal closure/cell death [37–39]. Therefore, H_2O_2 elevation and subsequent activation of Ca^{2+} channels are events occurring in SO_2-induced stomatal closure/cell death. Ca^{2+} influx from extracellular region results in $[Ca^{2+}]_{cyt}$ increase, and then $[Ca^{2+}]_{cyt}$ elevation mediates subsequent stomatal closure/cell death. These results suggested that H_2O_2 mediates SO_2-induced stomatal movement/cytotoxicity by targeting Ca^{2+} channels in the plasma membrane.

Briefly, environmental SO_2 has a remarkable effect on the size of the stomatal aperture. Exposure to SO_2 induced the overproduction of H_2O_2 in guard cells, as shown in other plant cells exposed to environmental challenges [40, 41]. Elevated H_2O_2 acts in conjunction with other factors to activate stomatal movement and cell death under SO_2 stress. Guard cells are a well-developed model system for characterizing early signal transduction mechanisms in plants. In this study, the dual role of H_2O_2 in plant cells in response to air pollutant was clearly displayed in Arabidopsis guard cells, which additionally broaden the role of stomatal guard cells in cytotoxicity study. Our results suggest that guard cells are a valuable model system for the study of cytotoxicity in plant cells.

CONCLUSION

Sulfur dioxide exposure caused an elevated H_2O_2 level in Arabidopsis guard cells. H_2O_2 elevation triggered by SO_2 mediated both stomatal closure and cell death via Ca^{2+} signaling. Intracellular Ca^{2+} increase is necessary for stomatal closure and cell death in Arabidopsis guard cells in response to SO_2. H_2O_2 production by NADPH oxidase plays a critical role in SO_2 toxicity; however NADPH oxidase activation is sufficient but not necessary for SO_2-triggered stomatal closure. Both stomatal closing and stomatal opening inhibition evoked by SO_2 lead to a decline in stomatal aperture, protecting the leaf against further entry of the pollutant but also curtailing photosynthesis. Cell death evoked by high concentrations of SO_2 indicated the cytotoxicity of SO_2, but might provide appropriate protection by reducing ROS production inArabidopsis plants.

ACKNOWLEDGMENTS

This study was supported by the National Natural Science Foundation of China (Grant nos. 30870454, 30470318, and 31371868), Research Fund for the Doctoral Program of Higher Education of China (Grant no. 20070108007), and Shanxi Scholarship Council of China (Grant no. 2009022).

REFERENCES

1. N. M. Darrall, "The effect of air pollutants on physiological processes in plants," Plant, Cell & Environment, vol. 12, no. 1, pp. 1–30, 1989.
2. T. Hogetsu and M. Shishikura, "Effects of sulfur dioxide and ozone on intact leaves and isolated mesophyll cells of groundnut plants (Arachis hypogaea L.)," Journal of Plant Research, vol. 107, no. 3, pp. 229–235, 1994.
3. M. Noji, M. Saito, M. Nakamura, M. Aono, H. Saji, and K. Saito, "Cysteine synthase overexpression in tobacco confers tolerance to sulfur-containing environmental pollutants," Plant Physiology, vol. 126, no. 3, pp. 973–980, 2001.
4. R. Rakwal, G. K. Agrawal, A. Kubo et al., "Defense/stress responses elicited in rice seedlings exposed to the gaseous air pollutant sulfur dioxide," Environmental and Experimental Botany, vol. 49, no. 3, pp. 223–235, 2003.
5. H. Rennenberg and C. Herschbach, "Responses of plants to atmospheric sulphur," inPlant Response to Air Pollution, M. Yunus and M. Iabal, Eds., pp. 285–294, John Wiley & Sons, Chichester, UK, 1996.
6. H. Pfanz and U. Heber, "Buffer capacities of leaves, leaf cells, and leaf cell organelles in relation to fluxes of potentially acidic gases," Plant Physiology, vol. 81, pp. 597–602, 1986.

7. K. Asada, "Formation and scavenging of superoxides in chloroplasts, with relation to injury by sulfur dioxide," Research Report of the National Institute for Environmental Studies, vol. 11, pp. 165–179, 1980.
8. N. R. Madamanchi and R. G. Alscher, "Metabolic bases for differences in sensitivity of two pea cultivars to sulfur dioxide," Plant Physiology, vol. 97, no. 1, pp. 88–93, 1991
9. K. Tanaka, N. Kondo, and K. Sugahara, "Accumulation of hydrogen peroxide in chloroplasts of SO_2-fumigated spinach leaves," Plant and Cell Physiology, vol. 23, no. 6, pp. 999–1007, 1982.
10. R. Hänsch and R. R. Mendel, "Sulfite oxidation in plant peroxisomes," Photosynthesis Research, vol. 86, no. 3, pp. 337–343, 2005.
11. E. Giraud, A. Ivanova, C. S. Gordon, J. Whelan, and M. J. Considine, "Sulphur dioxide evokes a large scale reprogramming of the grape berry transcriptome associated with oxidative signalling and biotic defence responses," Plant, Cell and Environment, vol. 35, no. 2, pp. 405–417, 2012.
12. L. Li and H. Yi, "Differential expression of Arabidopsis defense-related genes in response to sulfur dioxide," Chemosphere, vol. 87, no. 7, pp. 718–724, 2012.
13. J. Dat, S. Vandenabeele, E. Vranová, M. van Montagu, D. Inzé, and F. van Breusegem, "Dual action of the active oxygen species during plant stress responses," Cellular and Molecular Life Sciences, vol. 57, no. 5, pp. 779–795, 2000.
14. P. Sharma, A. B. Jha, R. S. Dubey, and M. Pessarakli, "Reactive Oxygen species, oxidative damage, and antioxidative defense mechanism in plants under stressful conditions,"Journal of Botany, vol. 2012, Article ID 217037, 26 pages, 2012.
15. S. Neill, R. Desikan, and J. Hancock, "Hydrogen peroxide signalling," Current Opinion in Plant Biology, vol. 5, no. 5, pp. 388–395, 2002.
16. C. Laloi, K. Apel, and A. Danon, "Reactive oxygen signalling: the latest news," Current Opinion in Plant Biology, vol. 7, no. 3, pp. 323–328, 2004.
17. H. B. Shao, L. Y. Chu, Z. H. Lu, and C. M. Kang, "Primary antioxidant free radical scavenging and redox signaling pathways

in higher plant cells," International Journal of Biological Sciences, vol. 4, no. 1, pp. 8–14, 2008.

18. R. Mittler, "Oxidative stress, antioxidants and stress tolerance," Trends in Plant Science, vol. 7, no. 9, pp. 405–410, 2002.
19. S.-H. Hung, C.-W. Yu, and C. H. Lin, "Hydrogen peroxide functions as a stress signal in plants," Botanical Bulletin of Academia Sinica, vol. 46, no. 1, pp. 1–10, 2005.
20. L. Li, H. Yi, L. Wang, and X. Li, "Effects of sulfur dioxide on the morphological and physiological biochemical parameters in Arabidopsis thaliana plants," Journal of Agro-Environment Science, vol. 27, pp. 525–529, 2008.
21. A. M. Hetherington, "Guard cell signaling," Cell, vol. 107, no. 6, pp. 711–714, 2001.
22. X. Zhang, L. Zhang, F. C. Dong, J. F. Gao, D. W. Galbraith, and C.-P. Song, "Hydrogen peroxide is involved in abscisic acid-induced stomatal closure in Vicia faba," Plant Physiology, vol. 126, no. 4, pp. 1438–1448, 2001.
23. J. I. Schroeder, G. J. Allen, V. Hugouvieux, J. M. Kwak, and D. Waner, "Guard cell signal transduction," Annual Review of Plant Biology, vol. 52, pp. 627–658, 2001.
24. H. Yi, J. Yin, X. Liu, X. Jing, S. Fan, and H. Zhang, "Sulfur dioxide induced programmed cell death in Vicia guard cells," Ecotoxicology and Environmental Safety, vol. 78, pp. 281–286, 2012.
25. K. Apel and H. Hirt, "Reactive oxygen species: metabolism, oxidative stress, and signal transduction," Annual Review of Plant Biology, vol. 55, pp. 373–399, 2004.
26. R. Desikan, M.-K. Cheung, A. Clarke et al., "Hydrogen peroxide is a common signal for darkness- and ABA-induced stomatal closure in Pisum sativum," Functional Plant Biology, vol. 31, no. 9, pp. 913–920, 2004.
27. N. Suzuki, S. Koussevitzky, R. Mittler, and G. Miller, "ROS and redox signalling in the response of plants to abiotic stress," Plant, Cell and Environment, vol. 35, no. 2, pp. 259–270, 2012.

28. T. Fukao and J. Bailey-Serres, "Plant responses to hypoxia—is survival a balancing act?"Trends in Plant Science, vol. 9, no. 9, pp. 449–456, 2004.
29. R. Mittler, S. Vanderauwera, M. Gollery, and F. van Breusegem, "Reactive oxygen gene network of plants," Trends in Plant Science, vol. 9, no. 10, pp. 490–498, 2004.
30. Y. Song, Y. Miao, and C.-P. Song, "Behind the scenes: the roles of reactive oxygen species in guard cells," New Phytologist, vol. 201, no. 4, pp. 1121–1140, 2014.
31. Z.-M. Pel, Y. Murata, G. Benning, et al., "Calcium channels activated by hydrogen peroxide mediate abscisic acid signalling in guard cells," Nature, vol. 406, no. 6797, pp. 731–734, 2000.
32. V. D. Petrov and F. van Breusegem, "Hydrogen peroxide—a central hub for information flow in plant cells," AoB Plants, vol. 12, no. 1, Article ID pls014, 2012.
33. M. A. Torres and J. L. Dangl, "Functions of the respiratory burst oxidase in biotic interactions, abiotic stress and development," Current Opinion in Plant Biology, vol. 8, no. 4, pp. 397–403, 2005. View at Publisher · View at Google Scholar · View at Scopus
34. R. Mahalingam and N. Fedoroff, "Stress response, cell death and signalling: the many faces of reactive oxygen species," Physiologia Plantarum, vol. 119, no. 1, pp. 56–68, 2003.
35. I. C. Mori and J. I. Schroeder, "Reactive oxygen species activation of plant Ca^{2+}channels. A signaling mechanism in polar growth, hormone transduction, stress signaling, and hypothetically mechanotransduction," Plant Physiology, vol. 135, no. 2, pp. 702–708, 2004.
36. D. B. Kiselevsky, Y. E. Kuznetsova, L. A. Vasil'ev et al., "Effect of Ca^{2+} on programmed death of guard and epidermal cells of pea leaves," Biochemistry, vol. 75, no. 5, pp. 614–622, 2010.
37. R. Errakhi, A. Dauphin, P. Meimoun et al., "An early Ca^{2+} influx is a prerequisite to thaxtomin A-induced cell death in Arabidopsis thaliana cells," Journal of Experimental Botany, vol. 59, no. 15, pp. 4259–4270, 2008.

38. S. Orrenius, B. Zhivotovsky, and P. Nicotera, "Regulation of cell death: the calcium-apoptosis link," Nature Reviews Molecular Cell Biology, vol. 4, no. 7, pp. 552–565, 2003.
39. F. Van Breusegem and J. F. Dat, "Reactive oxygen species in plant cell death," Plant Physiology, vol. 141, no. 2, pp. 384–390, 2006.
40. M. C. de Pinto, V. Locato, and L. de Gara, "Redox regulation in plant programmed cell death," Plant, Cell and Environment, vol. 35, no. 2, pp. 234–244, 2012.
41. T. Pfannschmidt, K. Bräutigam, R. Wagner, et al., "Potential regulation of gene expression in photosynthetic cells by redox and energy state: approaches towards better understanding," Annals of Botany, vol. 103, no. 4, pp. 599–607, 2009.

Chapter 8

Biotechnological Approaches to Study Plant Responses to Stress

Rosa M. Pérez-Clemente, Vicente Vives, Sara I. Zandalinas, María F. López-Climent, Valeria Muñoz, and Aurelio Gómez-Cadenas

Department of Agricultural Sciences, Universitat Jaume I, Campus Riu Sec, 12071 Castelló de la Plana, Spain

ABSTRACT

Multiple biotic and abiotic environmental stress factors affect negatively various aspects of plant growth, development, and crop productivity. Plants, as sessile organisms, have developed, in the course of their evolution, efficient strategies of response to avoid, tolerate, or adapt to different types of stress situations. The diverse stress factors that plants have to face often activate similar cell signaling pathways and cellular responses, such as the production of stress proteins, upregulation of the antioxidant machinery, and accumulation of compatible solutes. Over the last few decades advances in plant physiology, genetics, and molecular biology have greatly improved our understanding of plant responses to abiotic stress conditions. In this paper, recent progresses on systematic analyses of plant responses to stress including genomics, proteomics, metabolomics, and transgenic-based approaches are summarized.

INTRODUCTION

Plants and animals share some response mechanisms to unfavorable environmental conditions; however, plants, being sessile organisms, have developed, in the course of their evolution, highly sophisticated and efficient strategies of response to cope with and adapt to different types of abiotic and biotic stress imposed by the frequently adverse environment.

Stress can be understood as a stimulus or influence that is outside the normal range of homeostatic control in a given organism: if a stress tolerance is exceeded, mechanisms are activated at molecular, biochemical, physiological, and morphological levels; once stress is controlled, a new physiological state is established, and homeostasis is reestablished. When the stress is retired, the plant may return to the original state or to a new physiological situation [1].

In the last years, and because of the great interest for both basic and applied research, there has been an important progress in the understanding of the mechanisms and processes underlying abiotic stress adaptation and defense in different plant species [1, 2]. The sensing of biotic or abiotic stress conditions induces signaling cascades that activate ion channels, kinase cascades, production of reactive oxygen species (ROS), accumulation of hormones such as salicylic acid, ethylene, jasmonic acid, and abscisic acid. These signals ultimately induce expression of specific subsets of defense genes that lead to the assembly of the overall defense reaction [3].

The emergence of the novel "omics" technologies, such as genomics, proteomics, and metabolomics, is now allowing researchers to identify the genetic behind plant stress responses (Figure 1). These omics technologies enable a direct and unbiased monitoring of the factors affecting plant growth and development and provide the data that can be directly used to investigate the complex interplay between the plant, its metabolism, and also the stress caused by the environment or the biological threats (insects, fungi, or other pathogens). Plant responses to stress are mediated via profound changes in gene expression which result in changes in composition of plant transcriptome, proteome, and metabolome [4].

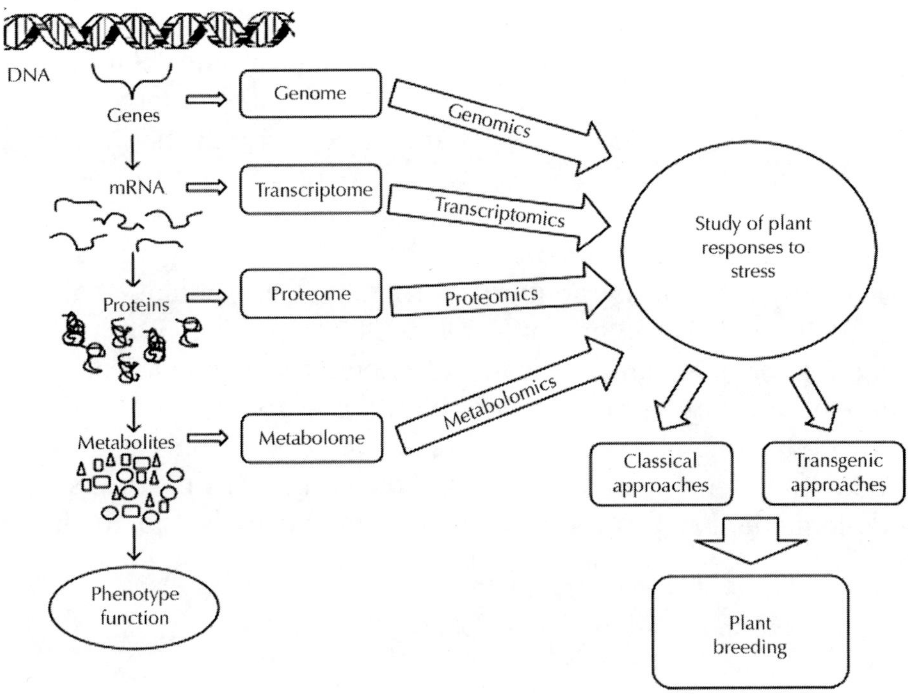

Figure 1. Plant response to abiotic stress factors. Genomics, transcriptomics, proteomics, and metabolomics have enabled active analyses of regulatory networks that control abiotic stress responses. Such analyses increase our knowledge on plant responses and adaptation to stress conditions and allow improving plant breeding.

In this work, the main biotechnological approaches to study plant responses to stress are reviewed.

GENOMICS

A gene by gene approach has been typically used to understand its function. In Table 1, some of the genes involved in plant responses to stress are listed. Functional genomics allows large-scale gene function analysis with high throughput technology and incorporates interaction of gene products at cellular and organism level. The information coming from sequencing programs is providing enormous input about genes to be analyzed. The availability of many plant genomes nowadays (reviewed in [5, 6]) facilitates studying the function of genes on a genomewide scale. The lack of information from other plant genomes will also be compensated in part by the availability of large collection of expressed sequence tags (ESTs) and cDNA sequences [7]. The basic interest behind these EST projects is to identify genes responsible for critical functions. ESTs, cDNA libraries, microarray, and serial analysis of gene expression (SAGE) are used to analyze global gene expression profiles in a functional genomics program. Large mutant collections are tools that complement large-scale expression studies. Gene identification through physical and chemical mutagens has become amenable for large-scale analysis with the availability of markers [8], but gene tagging is more promising for functional analysis on a wider scale. Moreover, the understanding of the complexity of stress signaling and plant adaptive processes would require the analysis of the function of numerous genes involved in stress response.

Table 1. Genes involved in plant responses to stress

	Stress	Reference
14.3.3 gene family (GF14b, GF14c)	Salinity, drought, fungal	[9]
MAPK	Abiotic and biotic stresses	[10]
MEKK1 and ANP1	Environmental stress	[11, 12]
MPK3, MPK4 and MPK6	Abiotic stress (pathogens) and oxidative stress	[10, 13]

CBF/DREB families (CBF1, CBF2, DREB2A)	Drought, cold, salinity	[47, 82]
HVA1	Salinity and drought	[83]
Glycerol-3-phosphate acyltransferase gene	Cold	[84]
ICS	Pathogens, UV light	[85]
LOX	Wounding, drought, and pathogens	[86, 87]
bZIPs family (e.g., ABF1, ABF2)	Drought, temperature, salinity	[24–33]
WRKY family (AtWRKY2, AtWRKY6, AtWRKY18)	Pathogens, wounding, salinity, temperature, drought, oxidative stress	[37]
ATAF	Wounding, drought, salinity, cold, pathogens	[88]

Numerous investigations show that plant defense response genes are transcriptionally activated by pathogens and also by different types of abiotic stress. It has been described that the induction of specific defense genes, in the response against certain pathogens, is dependent on specific environmental conditions, suggesting the existence of a complex signaling network that allows the plant to recognize and protect itself against pathogens and environmental stress [3]. Similar induction patterns of members of the 14.3.3 gene family (GF14b and GF14c) by abiotic and biotic stresses such as salinity, drought, ABA, and fungal inoculation have been documented in rice [9]. The rice GF14 genes contain cis-elements in their promoter regions that are responsive to abiotic stress and pathogen attack. The 14-3-3s family genes are also subject to the regulation by certain transcript factors [9].

On the other hand, kinase cascades of the mitogen-activated protein kinase (MAPK) class play a remarkably important role in plant signaling of a variety of abiotic and biotic stresses, and it is an essential step in the establishment of resistance to pathogens [10]. It has been described that in Arabidopsis MEKK1 and ANP1 act in the environmental stress response [11, 12], and MPK3, MPK4, and MPK6, are activated by a diversity of stimuli including abiotic stresses, pathogens, and oxidative stress [13].

Elucidating the molecular mechanism that mediates the complex stress responses in plants system is an important step to develop improved variety of stress tolerant crops. Many crop traits are quantitative, complex, and controlled by multiple interacting genes. Recent progress in molecular biology provides the tools to study the genetical make-up of plants, which allows us to unravel the inheritance of all traits whether they are controlled by single genes or many genes acting together, known as the quantitative trait loci (QTL). The molecular marker technologies available since the 1980s allows dissecting the variation in traits. With the progress of QTL mapping, new breeding approaches such as marker-assisted selection and breeding by design have emerged [14]. Advances in plant genomics research have opened up new perspectives and opportunities for improving crop plants and their productivity. The genomics technologies have been found useful in deciphering the multigenicity of biotic and abiotic plant stress responses through genome sequences, stress-specific cell and tissue transcript collections, protein and metabolite profiles and their dynamic changes, protein interactions, and mutant screens.

PROTEOMICS

The adaptation of plants to biotic or abiotic stress conditions is mediated through deep changes in gene expression which result in changes in composition of plant transcriptome, proteome, and metabolome. Since proteins are directly involved in plant stress response, proteomics studies can significantly contribute to elucidate the possible relationships between protein abundance and plant stress acclimation. Several studies [15] have already proven that the changes in gene expression at transcript level do not often correspond with the changes at protein level. The investigation of changes in plant proteome is highly important since proteins, unlike transcripts, are direct effectors of plant stress response. Proteins not only include enzymes catalyzing changes in metabolite levels, but also include components of transcription and translation machinery (Table 2).

Table 2. Proteins and enzymes involved in plant responses to stress

	Stress	Reference
ERF family	Cold, drought, pathogen infection, wounding, ET, SA, and JA	[19]
bZIPs family (e.g., ABF1, ABF2)	Drought, temperature, salt	[24–33]
WRKY	Pathogens, wounding, salinity, temperature, drought, oxidative stress	[36–39]
MYB family (AtMYB15, AtMYB30, AtMYB33 AtMYB60, AtMYB96, AtMYB101 AtMYB15, and AtMYB108)	Biotic and abiotic stress (pathogens, drought, cold)	[40–46]
ABF	Drought	[79]
NAC	Drought, salinity, cold	[22, 23]
MYC	Environmental stresses	[47]
LEA family (PMA 80, PMA 1959)	Salinity and drought	[71]
Heat shock proteins	Temperatures	[74, 75]
LOX family (e.g., LOX1)	Wounding, drought, and pathogens	[86, 87]
Glutathione peroxidase, superoxide dismutase, ascorbate peroxidases, and glutathione reductases	Oxidative stress	[68, 70]

In the last years, there has been an important progress in the knowledge of several families of plant transcription factors linked to plant stress responses, such as responses to ultraviolet light, wounding, anaerobic stress, and pathogens [16]. The most important ones are as follows.

- The ethylene-responsive-element-binding factors (ERFs). This protein family has been linked to a wide range of stresses; the RNA levels of specific ERF genes are regulated by cold, drought, pathogen infection, wounding or treatment with ethylene, SA or JA [17]. ERF proteins are shown to function as either activators or repressors of transcription, which is of great relevance in all processes related to plant development and its responses to adverse growing conditions due to both biotic and abiotic factors [18]. It has been reported that ERF proteins from one plant species function in other plant species, enhancing their potential utility in increasing the stress tolerance of plants [19, 20]. However, constitutive overexpression of ERF genes generally causes deleterious effects. To overcome this problem, the use of stress-inducible promoters to control the expression of the ERF genes has been successfully used (reviewed in [21]).

- NAC proteins are plant-specific transcription factors having a variety of important functions not only in plant development but also in abiotic stress tolerance [22]. NAC domain-containing proteins represent one of the largest TF families, firstly identified in model plants as Arabidopsis and rice but also recently characterized in woody fruit species [23].
- Another important family of transcription factors is the called "basic-domain leucine-zipper (bZIP)" which are regulators of important plant processes such as organ and tissue differentiation [24], cell elongation [25], nitrogen/carbon balance control [26], pathogen defense [27], energy metabolism [28], unfolded protein response [29], hormone and sugar signaling [30], light response [31], osmotic control [32], and seed storage protein gene regulation [33]. One class of bZIP proteins that is linked to stress responses comprises the TGA/octopine synthase (ocs)-element-binding factor (OBF) proteins. These bind to the activation sequence-1 (as-1)/ocs element, which regulates the expression of some stress-responsive genes [34]. A major advance was the discovery that TGA/OBF family members interact with nonexpressor of PR1 (NPR1), a key component in the SA defense signaling pathway [35].
- WRKY proteins are a family of transcription factors that are unique to plants specific WRKY family members show enhanced expression and/or DNA-binding activity following induction by a range of pathogens, defense signals, and wounding (reviewed in [36]). Significant progress has been made in the past years in identifying target genes for WRKY factors. WRKY proteins bind to the W box, which is found in the promoters of many plant defense genes [37]. WRKY proteins also regulate the expression

of regulatory genes such as receptor protein kinases [38]. Positive and negative regulation of WRKY promoters by specific WRKY proteins has been observed, and the promoters of many of the pathogen- and/or SA-regulated At WRK genes are rich in W boxes [39].

- MYB proteins are key factors in regulatory networks controlling development [40], metabolism [41], and responses to biotic and abiotic stresses [42]. Since the Arabidopsis genome sequence was published, some years ago, an important amount of data has accumulated on the roles of MYB transcription factors in plants and some members of this family are involved in these responses. Therefore, AtMYB30 encodes an activator of the hypersensitive cell death program in response to pathogen attack [43]; AtMYB96 acts through the ABA signaling cascade to regulate water stress and disease resistance [44]. AtMYB33 and AtMYB101 are involved in ABA-mediated responses to environmental signals. AtMYB15 is also involved in cold stress tolerance [45]. AtMYB108 in both biotic and abiotic stress responses [46]. The elucidation of MYB protein function and regulation that is possible in Arabidopsis will allow predicting the contributions of MYB proteins to the responses to biotic and abiotic stress conditions in other plant species.
- (vi)MYC proteins are involved in the response of plants to unfavorable environmental conditions. This transcription factor family plays a role in the induction of apoptosis, important in the hypersensitive cell death program in response to pathogen attack. Another putative MYC target is the ornithine decarboxylase gene, involved in polyamines synthesis. On the other hand, MYC proteins activate the major ABA-dependent stress response (reviewed in [47]).

Therefore, studies of plant reaction upon stress conditions at protein level can significantly contribute to our understanding of physiological mechanisms underlying plant stress tolerance. Proteomics studies could thus lead to identification of potential protein markers whose changes in abundance can be associated with quantitative changes in some physiological parameters related to stress tolerance (reviewed in [16]).

METABOLOMICS

The possibility of monitoring a complete set of metabolites could largely improve the understanding of many physiological plant processes. This systematic study, defined as "metabolomics," is intended to provide an integrated view of the functional status of an organism. Besides its use as a breeding or selection tool, metabolomics techniques have also been used to evaluate stress responses in barley [48], Citrus [49],Medicago truncatula [50], and Arabidopsis thaliana [51]. Some of the metabolites that have been involved in the plant responses to stress are listed in Table 3.

Table 3. Metabolites and hormones involved in plant responses to stress

	Stress	Reference
Abscisic acid, jasmonic acid, salicylic acid, polyamines, and others	Drought, salinity, cold	[3, 58, 65–67]
Proline, glycine-betaine, and other compatible osmolytes	Environmental stresses: drought, salinity, osmotic	[62–64]
Phytoalexins	Microbial pathogens	[89]
Terpenes	Toxins and pathogens	[89]
Phenolic compounds (coumarin, lignin, flavonoids, tannins, isoflavonoids)	Pathogens, oxidative stress, UV light	[89, 90]
Alkaloids	Pathogens (predators)	[87]
Unsaturated fatty acids	Environmental stresses	[73]
ROS, malondialdehyde	Biotic and abiotic stresses	[3, 58, 68]
Phytochelatins and metallothioneins	Heavy metal intoxication	[76, 77]

This technological tool, recently developed, includes different approaches, namely, targeted analysis, metabolic fingerprinting, and metabolite profiling.

Targeted analysis is the most developed analytical approach in metabolomics [49]. It is used to measure the concentration of a limited number of known metabolites precisely, by using either gas chromatography (GC) or liquid chromatography (LC) coupled to mass spectrometry (MS) or nuclear magnetic resonance spectroscopy (NMR).

Other approaches using high throughput metabolite analysis focus on a subset of useful information while avoiding the difficulties of comprehensive metabolite characterization; metabolic fingerprinting uses signals from hundreds to thousands of metabolites for rapid sample classification via statistical analysis [52]. In the last years, metabolite profiling attempts to identify and quantify a specific class or classes of chemically related metabolites that often share chemical properties that facilitate simultaneous analysis (reviewed in [53]).

The metabolome represents the downstream result of gene expression and is closer to phenotype than transcript expression or proteins. Extensive knowledge on metabolic flows could allow assessment of genotypic or phenotypic differences between plant species or among genotypes exhibiting different tolerance to some biotic or abiotic stresses. In addition, target metabolites have been analyzed as nutritional and/or agronomical biomarkers to classify different crop cultivars or to optimize growth conditions [54].

In contrast to high throughput methodology for the analysis of DNA, RNA, and proteins, current strategies for metabolite characterization still face significant obstacles. These challenges are largely caused by the high degree of chemical diversity among metabolite pools as well as the complexity of spatial and temporal distribution within living tissues. Plant metabolomics methodology and instrumentation are being developed at a rapid pace to address these analytical challenges [55].

Like other functional genomics research, metabolomics generates large amounts of data. Handling, processing, and analyzing this data is a clear challenge for researchers and requires specialized mathematical, statistical, and bioinformatic tools [56]. Further developments in this area require improvements in both analytical science and bioinformatics. Development of new analytical techniques is largely focused on increasing resolution and comprehensiveness, increasing speed and throughput of analytical assays and equipment miniaturization.

TRANSGENIC APPROACHES: FROM THE STUDY OF STRESS TOLERANCE MECHANISMS IN PLANTS TO CROP GENETIC IMPROVEMENT

Use of modern molecular biology tools for elucidating the control mechanisms of stress tolerance and for engineering stress tolerant plants is based on the expression of specific stress-related genes. To date, successes in genetic improvement of environmental stress resistance have involved manipulation of a single or a few genes involved in signaling/regulatory pathways or that encode enzymes involved in these pathways [57].

The plant hormone abscisic acid (ABA) regulates the adaptive response of plants to environmental stresses such as drought, salinity, and chilling via diverse physiological and developmental processes [58]. The ABA biosynthetic pathway has been deeply studied, and many of the key enzymes involved in ABA synthesis have been used in transgenic plants to improve abiotic stress tolerance [59]. Transgenic plants overexpressing the genes involved in ABA synthesis showed increased tolerance to drought and salinity stress [59]. Similarly, many studies have illustrated the potential of manipulating CBF/DREB genes to confer improved drought tolerance [60].

Another mechanism involved in plant protection to osmotic stress associated to many abiotic stresses such as drought and salinity implies the accumulation of compatible solutes involved in avoiding oxidative damage and chaperoning through direct stabilization of membranes and/or proteins [61]. Many genes involved in the synthesis of these osmoprotectants have been explored for their potential in engineering plant abiotic stress tolerance [61]. The amino acid proline is known to occur widely in higher plants and normally accumulates in large quantities in response to environmental stresses [62]. The osmoprotectant role of proline has been verified in some crops by overexpressing genes involved in proline synthesis [63]. The results of transgenic modifications of biosynthetic and metabolic pathways in most of the previously mentioned cases indicate that higher stress tolerance and the accumulation of compatible solutes may also protect plants against damage by scavenging of reactive oxygen species (ROS) and by their chaperone-like activities in maintaining protein structures and functions [64].

Polyamines, being polycationic compounds of low molecular weight, are involved in many cellular processes, such as replication, transcription, translation, membrane stabilization, enzyme activity modulation, plant growth, and development [65]. It has been reported that stress results in an accumulation of free or conjugated polyamines, indicating that polyamine biosynthesis might serve as an integral component of plant response to stress [66, 67].

Polyamines metabolic pathways are regulated by a limited number of key enzymes, among them ornithine decarboxylase (ODC) and arginine decarboxylase (ADC). Transgenic plants overexpressing ADC gene showed increase in biomass and better performance under salt stress conditions. It has also been described that genetic transformation with genes encoding ADC improved environmental stress tolerance in various plant species [66].

A common factor among most stresses is the active production of reactive oxygen species [2]. ROS are not only toxic to cells but also play an important role as signaling molecules. Under normal growth conditions, there is equilibrium between the production and the scavenging of ROS, but abiotic stress factors may disturb this equilibrium, leading to a sudden increase in intracellular levels of ROS.

In order to control the level of ROS and protect the cells from oxidative injury, plants have developed a complex antioxidant defense system to scavenge them [68]. These antioxidant systems include various enzymes and nonenzymatic metabolites that may also play a significant role in ROS signaling in plants. A number of transgenic improvements for abiotic stress tolerance have been achieved through detoxification strategy [69]. These include transgenic plants over expressing enzymes involved in oxidative protection, such as glutathione peroxidase, superoxide dismutase, ascorbate peroxidases, and glutathione reductases [70].

LEA proteins, including several groups of high molecular weight, accumulate in response to different environmental stresses. It has been reported that constitutive overexpression of the HVA1, a group 3 LEA protein from barley, conferred tolerance to soil water deficit and salt stress in transgenic rice plants [71]. It has also been reported that plants expressing a wheat LEA group 2 protein (PMA80) gene or the wheat LEA group 1protein (PMA1959) gene resulted in increased tolerance to dehydration and salt stresses [69].

An important strategy for achieving greater tolerance to abiotic stress is to help plants to reestablish homeostasis under stressful environments, restoring both ionic and osmotic homeostasis. This is a major approach to improve salt tolerance in plants through genetic engineering, where the target is to achieve Na^+excretion out of the root, or their storage in the vacuole [72].

Transgenic approaches also aim to improve photosynthesis under abiotic stress conditions through changes in the lipid biochemistry of the membranes. Genetically engineered plants overexpressing chloroplast glycerol-3-phosphate acyltransferase gene (involved in phosphatidyl glycerol fatty acid desaturation) showed an increase in the number of unsaturated fatty acids and a corresponding decrease in the chilling sensitivity [73].

The heat shock response is a highly conserved biological response, occurring in all organisms in response to heat or other toxic agent exposures [74]. Genetic engineering for increased thermotolerance by enhancing heat shock protein synthesis in plants has been achieved in a number of plant species. Some authors have reported the positive correlation between the levels of heat shock proteins and stress tolerance (reviewed in [75]).

A special case of study is the heavy metal contamination. In spite of the natural occurrence of heavy metals as rare elements, diverse anthropogenic practices have contributed to spread them in the environment. Plants have developed mechanisms that can protect cells from heavy metal cytotoxicity, as the cytosolic detoxification by binding to the metal-binding molecules as phytochelatins, and metallothioneins which play an important role in heavy metal detoxification and homeostasis of intracellular metal ions in plant tissues. Overexpression of phytochelatin synthase in Arabidopsis leads to enhanced arsenic tolerance but surprisingly to cadmium hypersensitivity [76]. Therefore, new approaches could contribute to uncovering the complexity of plant tolerance to heavy metal stress [77].

The transcription factors activate cascades of genes that act together in enhancing tolerance towards multiple stresses as indicated before.

On the other hand, some stress responsive genes may share the same transcription factors, as indicated by the significant overlap of the gene expression profiles that are induced in response to different stresses [37]. Transcriptional activation of stress-induced genes has been possible in transgenic plants over expressing one or more transcription factors that recognize promoter regulatory elements of these genes [75, 78]. Two families, bZIP and MYB, are involved in ABA signaling and its gene activation. Introduction of transcription factors in the ABA signaling pathway can also be a mechanism of genetic improvement of plant stress tolerance. Constitutive expression of ABF3 or ABF4 demonstrated enhanced drought tolerance in Arabidopsis, with altered expression of ABA/stress-responsive genes, for example, rd29B, rab18, ABI1, and ABI2 [79].

It is important to point that genetic modification of higher plants by introducing DNA into their cells is a highly complex process. Practically any plant transformation experiment relies at some point on cell and tissue culture. Although the development transformation methods that avoid plant tissue culture have been described for Arabidopsis and have been extended to a few crops, the ability to regenerate plants from isolated cells or tissues in vitro is needed for most plant transformation systems. Not all plant tissue is suited to every plant transformation method, and not all plant species can be regenerated by every method [80]. There is, therefore, a need to find both a suitable plant tissue culture/regeneration regime and a compatible plant transformation methodology [81].

CONCLUSIONS

To understand how plants respond to stress, it must be considered that they are subjected to a combination of adverse conditions. This preliminary consideration is essential to understand the performance of plants under stress and also to identify strategies to improve stress tolerance.

The integration of the omics approaches is likely to enable researchers to reconstruct the whole cascade of cellular events leading to rapid responses and adaptation to the various abiotic stimuli. A well-focused approach combining molecular, physiological, and metabolic aspects of plant stress tolerance is required to increase knowledge on the effects of gene expression and to understand whole plant phenotype under stress. A better understanding of the underlying physiological processes in response to different abiotic stresses can drive the selection of the appropriate promoter or transcription factor to be used for transformation.

In addition, the use of genetic and genomic analysis to identify DNA molecular markers associated to stress resistance can facilitate breeding strategies for crop improvement. This approach is particularly useful when target characters are controlled by several genes, as in the case of abiotic stress tolerance. These omics approaches could be combined with the potential to map different QTLs contributing to a given agronomical trait and to identify linked molecular markers. This will open the possibility to transfer simultaneously several QTLs and to pyramid QTLs for several agronomical traits in one improved cultivar.

ACKNOWLEDGMENTS

This work was supported by the Spanish Ministerio de Economía y Competitividad (MINECO) and Universitat Jaume I/Fundació Bancaixa through Grants no. AGL2010-22195-C03-01/AGR and P11B2009-01, respectively.

REFERENCES

1. S. Fraire-Velázquez, R. Rodríguez-Guerra, and L. Sánchez-Calderón, "Abiotic and biotic stress response crosstalk in plants," in Abiotic Stress Response in Plants—Physiological, Biochemical and Genetic Perspectives, A. Shanker, Ed., pp. 3–26, InTech, Rijeka, Croatia, 2011.
2. T. Hirayama and K. Shinozaki, "Research on plant abiotic stress responses in the post-genome era: past, present and future," Plant Journal, vol. 61, no. 6, pp. 1041–1052, 2010.
3. P. Jaspers and J. Kangasjärvi, "Reactive oxygen species in abiotic stress signaling," Physiologia Plantarum, vol. 138, no. 4, pp. 405–413, 2010.

4. F. Pérez-Alfocea, M. E. Ghanem, A. Gómez-Cadenas, and I. Dodd, "Omics of root-to-shoot signaling under salt stress and water deficit," OMICs A Journal of Integrative Biology, vol. 15, no. 12, pp. 893–901, 2011.
5. C. Feuillet, J. E. Leach, J. Rogers, P. S. Schnable, and K. Eversole, "Crop genome sequencing: lessons and rationales," Trends in Plant Science, vol. 16, no. 2, pp. 77–88, 2010.
6. P. S. G. Chain, D. V. Grafham, R. S. Fulton et al., "Genome project standards in a new era of sequencing," Science, vol. 326, no. 5950, pp. 236–237, 2009.
7. M. C. Marques, H. Alonso-Cantabrana, J. Forment et al., "A new set of ESTs and cDNA clones from full-length and normalized libraries for gene discovery and functional characterization in citrus,"BMC Genomics, vol. 10, article 428, 2009.
8. W. Lukowitz, C. S. Gillmor, and W. R. Scheible, "Positional cloning in Arabidopsis. Why it feels good to have a genome initiative working for you," Plant Physiology, vol. 123, no. 3, pp. 795–805, 2000.
9. F. Chen, Q. Li, L. Sun, and Z. He, "The rice 14-3-3 gene family and its involvement in responses to biotic and abiotic stress," DNA Research, vol. 13, no. 2, pp. 53–63, 2006.
10. A. Pitzschke, A. Schikora, and H. Hirt, "MAPK cascade signalling networks in plant defence," Current Opinion in Plant Biology, vol. 12, no. 4, pp. 421–426, 2009.
11. H. Nakagami, H. Soukupová, A. Schikora, V. Žárský, and H. Hirt, "A mitogen-activated protein kinase kinase kinase mediates reactive oxygen species homeostasis in Arabidopsis," Journal of Biological Chemistry, vol. 281, no. 50, pp. 38697–38704, 2006.
12. M. C. Suarez-Rodriguez, L. Adams-Phillips, Y. Liu et al., "MEKK1 is required for flg22-induced MPK4 activation in Arabidopsis plants," Plant Physiology, vol. 143, no. 2, pp. 661–669, 2007.
13. J. L. Qiu, L. Zhou, B. W. Yun et al., "Arabidopsis mitogen-activated protein kinase kinases MKK1 and MKK2 have overlapping functions in defense signaling mediated by MEKK1, MPK4, and MKS1,"Plant Physiology, vol. 148, no. 1, pp. 212–222, 2008. · ·

14. J. D. Peleman and J. R. van der Voort, "Breeding by design," Trends in Plant Science, vol. 8, no. 7, pp. 330–334, 2003.
15. M. B. Bogeat-Triboulot, M. Brosché, J. Renaut et al., "Gradual soil water depletion results in reversible changes of gene expression, protein profiles, ecophysiology, and growth performance in Populus euphratica, a poplar growing in arid regions," Plant Physiology, vol. 143, no. 2, pp. 876–892, 2007.
16. K. Kosová, P. Vítámvás, I. T. Prášil, and J. Renaut, "Plant proteome changes under abiotic stress—contribution of proteomics studies to understanding plant stress response," Journal of Proteomics, vol. 74, no. 8, pp. 1301–1322, 2011.
17. L. Oñate-Sánchez and K. B. Singh, "Identification of Arabidopsis ethylene-responsive element binding factors with distinct induction kinetics after pathogen infection," Plant Physiology, vol. 128, no. 4, pp. 1313–1322, 2002.
18. S. Y. Fujimoto, M. Ohta, A. Usui, H. Shinshi, and M. Ohme-Takagi, "Arabidopsis ethylene-responsive element binding factors act as transcriptional activators or repressors of GCC box-mediated gene expression," Plant Cell, vol. 12, no. 3, pp. 393–404, 2000.·
19. T. Nakano, K. Suzuki, T. Fujimura, and H. Shinshi, "Genome-wide analysis of the ERF gene family inArabidopsis and rice," Plant Physiology, vol. 140, no. 2, pp. 411–432, 2006.
20. L. Wu, X. Chen, H. Ren et al., "ERF protein JERF1 that transcriptionally modulates the expression of abscisic acid biosynthesis-related gene enhances the tolerance under salinity and cold in tobacco,"Planta, vol. 226, no. 4, pp. 815–825, 2007.
21. J. Mizoi, K. Shinozaki, and K. Yamaguchi-Shinozaki, "AP2/ERF family transcription factors in plant abiotic stress responses," Biochimica et Biophysica Acta, vol. 1819, no. 2, pp. 86–96, 2012.
22. K. Nakashima, H. Takasaki, J. Mizoi, K. Shinozaki, and K. Yamaguchi-Shinozaki, "NAC transcription factors in plant biotic and abiotic stress responses," Biochimica et Biophysica Acta, vol. 1819, no. 2, pp. 97–103, 2011.
23. T. M. de Oliveira, L. C. Cidade, A. S. Gesteira, M. A. Coelho Filho, W. S. Soares Filho, and M. G. C. Costa, "Analysis of the NAC

transcription factor gene family in citrus reveals a novel member involved in multiple abiotic stress responses," Tree Genetics and Genomes, vol. 7, no. 6, pp. 1123–1134, 2011.

24. M. Abe, Y. Kobayashi, S. Yamamoto et al., "FD, a bZIP protein mediating signals from the floral pathway integrator FT at the shoot apex," Science, vol. 309, no. 5737, pp. 1052–1056, 2005.
25. L. G. Corrêa, D. M. Riaño-Pachón, C. G. Schrago, R. V. dos Santos, B. Mueller-Roeber, and M. Vincentz, "The role of bZIP transcription factors in green plant evolution: adaptive features emerging from four founder genes," PLoS ONE, vol. 3, no. 8, article e2944, 2008.
26. F. Weltmeier, A. Ehlert, C. S. Mayer et al., "Combinatorial control of Arabidopsis proline dehydrogenase transcription by specific heterodimerisation of bZIP transcription factors," EMBO Journal, vol. 25, no. 13, pp. 3133–3143, 2006.
27. H. Kaminaka, C. Näke, P. Epple et al., "bZIP10-LSD1 antagonism modulates basal defense and cell death in Arabidopsis following infection," EMBO Journal, vol. 25, no. 18, pp. 4400–4411, 2006.
28. E. Baena-González, F. Rolland, J. M. Thevelein, and J. Sheen, "A central integrator of transcription networks in plant stress and energy signalling," Nature, vol. 448, no. 7156, pp. 938–942, 2007.
29. J. X. Liu, R. Srivastava, P. Che, and S. H. Howell, "Salt stress responses in Arabidopsis utilize a signal transduction pathway related to endoplasmic reticulum stress signaling," Plant Journal, vol. 51, no. 5, pp. 897–909, 2007.
30. C. Nieva, P. K. Busk, E. Domínguez-Puigjaner et al., "Isolation and functional characterisation of two new bZIP maize regulators of the ABA responsive gene rab28," Plant Molecular Biology, vol. 58, no. 6, pp. 899–914, 2005.
31. B. Al-Sady, W. Ni, S. Kircher, E. Schäfer, and P. H. Quail, "Photoactivated phytochrome induces rapid PIF3 phosphorylation prior to proteasome-mediated degradation," Molecular Cell, vol. 23, no. 3, pp. 439–446, 2006.
32. R. Satoh, Y. Fujita, K. Nakashima, K. Shinozaki, and K. Yamaguchi-Shinozaki, "A novel subgroup of bZIP proteins functions as transcriptional activators in hypoosmolarity-responsive

expression of the ProDH gene in Arabidopsis," Plant and Cell Physiology, vol. 45, no. 3, pp. 309–317, 2004.

33. R. Alonso, L. Oñate-Sánehez, F. Weltmeier et al., "A Pivotal role of the basic leucine zipper transcription factor bZIP53 in the regulation of Arabidopsis seed maturation gene expression based on heterodimerization and protein complex formation," Plant Cell, vol. 21, no. 6, pp. 1747–1761, 2009.
34. B. Fode, T. Siemsen, C. Thurow, R. Weigel, and C. Gatz, "The Arabidopsis GRAS protein SCL14 interacts with class II TGA transcription factors and is essential for the activation of stress-inducible promoters," Plant Cell, vol. 20, no. 11, pp. 3122–3135, 2008.
35. G. Le Henanff, T. Heitz, P. Mestre, J. Mutterer, B. Walter, and J. Chong, "Characterization of Vitis vinifera NPR1 homologs involved in the regulation of pathogenesis-related gene expression," BMC Plant Biology, vol. 9, article 54, 2009.
36. P. J. Rushton, I. E. Somssich, P. Ringler, and Q. J. Shen, "WRKY transcription factors," Trends in Plant Science, vol. 15, no. 5, pp. 247–258, 2010.
37. D. Qiu, J. Xiao, W. Xie, H. Cheng, X. Li, and S. Wang, "Exploring transcriptional signalling mediated by OsWRKY13, a potential regulator of multiple physiological processes in rice," BMC Plant Biology, vol. 9, article 74, 2009.
38. J. Wan, X. C. Zhang, and G. Stacey, "Chitin signaling and plant disease resistance," Plant Signaling and Behavior, vol. 3, no. 10, pp. 831–833, 2008.
39. S. G. Kang, E. Park, and K. S. Do, "Identification of a pathogen-Induced glycine max transcription factor GmWRKY1," Plant Pathology Journal, vol. 25, no. 4, pp. 381–388, 2009.
40. A. Mandaokar and J. Browse, "MYB108 acts together with MYB24 to regulate jasmonate-mediated stamen maturation in Arabidopsis," Plant Physiology, vol. 149, no. 2, pp. 851–862, 2009.
41. L. Lepiniec, I. Debeaujon, J. M. Routaboul et al., "Genetics and biochemistry of seed flavonoids,"Annual Review of Plant Biology, vol. 57, pp. 405–430, 2006.
42. C. Dubos, R. Stracke, E. Grotewold, B. Weisshaar, C. Martin, and

L. Lepiniec, "MYB transcription factors in Arabidopsis," Trends in Plant Science, vol. 15, no. 10, pp. 573–581, 2010.

43. S. Raffaele, F. Vailleau, A. Léger et al., "A MYB transcription factor regulates very-long-chain fatty acid biosynthesis for activation of the hypersensitive cell death response in Arabidopsis," Plant Cell, vol. 20, no. 3, pp. 752–767, 2008.
44. P. J. Seo and C. M. Park, "MYB96-mediated abscisic acid signals induce pathogen resistance response by promoting salicylic acid biosynthesis in Arabidopsis," New Phytologist, vol. 186, no. 2, pp. 471–483, 2010.
45. M. Agarwal, Y. Hao, A. Kapoor et al., "A R2R3 type MYB transcription factor is involved in the cold regulation of CBF genes and in acquired freezing tolerance," Journal of Biological Chemistry, vol. 281, no. 49, pp. 37636–37645, 2006.
46. L. Zhang, G. Zhao, J. Jia, and X. Liu, "Molecular characterization of 60 isolated wheat MYB genes and analysis of their expression during abiotic stress," Journal of Experimental Botany, vol. 63, no. 1, pp. 203–214, 2011.
47. P. K. Agarwal and B. Jha, "Transcription factors in plants and ABA dependent and independent abiotic stress signalling," Biologia Plantarum, vol. 54, no. 2, pp. 201–212, 2010.
48. Widodo, J. H. Patterson, E. Newbigin, M. Tester, A. Bacic, and U. Roessner, "Metabolic responses to salt stress of barley (Hordeum vulgare L.) cultivars, Sahara and Clipper, which differ in salinity tolerance," Journal of Experimental Botany, vol. 60, no. 14, pp. 4089–4103, 2009.
49. J. D. Djoukeng, V. Arbona, R. Argamasilla, and A. Gomez-Cadenas, "Flavonoid profiling in leaves of citrus genotypes under different environmental situations," Journal of Agricultural and Food Chemistry, vol. 56, no. 23, pp. 11087–11097, 2008.
50. C. D. Broeckling, D. V. Huhman, M. A. Farag et al., "Metabolic profiling of Medicago truncatula cell cultures reveals the effects of biotic and abiotic elicitors on metabolism," Journal of Experimental Botany, vol. 56, no. 410, pp. 323–336, 2005.
51. A. Fukushima, M. Kusano, H. Redestig, M. Arita, and K. Saito, "Metabolomic correlation-network modules in Arabidopsis based

on a graph-clustering approach," BMC Systems Biology, vol. 5, article no. 1, 2011.

52. S. Chatterjee, S. Srivastava, A. Khalid et al., "Comprehensive metabolic fingerprinting of Withania somnifera leaf and root extracts," Phytochemistry, vol. 71, no. 10, pp. 1085–1094, 2010.
53. C. Seger and S. Sturm, "Analytical aspects of plant metabolite profiling platforms: current standings and future aims," Journal of Proteome Research, vol. 6, no. 2, pp. 480–497, 2007.
54. V. Arbona, R. Argamasilla, and A. Gómez-Cadenas, "Common and divergent physiological, hormonal and metabolic responses of Arabidopsis thaliana and Thellungiella halophila to water and salt stress," Journal of Plant Physiology, vol. 167, no. 16, pp. 1342–1350, 2010.
55. A. D. Hegeman, "Plant metabolomics-meeting the analytical challenges of comprehensive metabolite analysis," Briefings in Functional Genomics and Proteomics, vol. 9, no. 2, pp. 139–148, 2010.
56. V. Shulaev, "Metabolomics technology and bioinformatics," Briefings in Bioinformatics, vol. 7, no. 2, pp. 128–139, 2006.
57. M. C. Jewell, B. C. Campbell, and I. D. Godwin, "Transgenic plants for abiotic stress resistance," inTransgenic Crop Plants, C. Kole, et al., Ed., Springer, Heidelberg, Germany, 2010.·
58. V. Arbona and A. Gómez-Cadenas, "Hormonal modulation of citrus responses to flooding," Journal of Plant Growth Regulation, vol. 27, no. 3, pp. 241–250, 2008.
59. X. Ji, B. Dong, B. Shiran et al., "Control of abscisic acid catabolism and abscisic acid homeostasis is important for reproductive stage stress tolerance in cereals," Plant Physiology, vol. 156, no. 2, pp. 550–563, 2011.
60. Y. Xiong and S. Z. Fei, "Functional and phylogenetic analysis of a DREB/CBF-like gene in perennial ryegrass (Lolium perenne L.)," Planta, vol. 224, no. 4, pp. 878–888, 2006.
61. Y. Y. Zhang, Y. Li, T. Gao et al., "Arabidopsis SDIR1 enhances drought tolerance in crop plants,"Bioscience, Biotechnology and Biochemistry, vol. 72, no. 8, pp. 2251–2254, 2008.

62. M. Ashraf and M. R. Foolad, "Roles of glycine betaine and proline in improving plant abiotic stress resistance," Environmental and Experimental Botany, vol. 59, no. 2, pp. 206–216, 2007.
63. A. Hmida-Sayari, R. Gargouri-Bouzid, A. Bidani, L. Jaoua, A. Savouré, and S. Jaoua, "Overexpression of D1-pyrroline-5-carboxylate synthetase increases proline production and confers salt tolerance in transgenic potato plants," Plant Science, vol. 169, no. 4, pp. 746–752, 2005.
64. T. Umezawa, M. Fujita, Y. Fujita, K. Yamaguchi-Shinozaki, and K. Shinozaki, "Engineering drought tolerance in plants: discovering and tailoring genes to unlock the future," Current Opinion in Biotechnology, vol. 17, no. 2, pp. 113–122, 2006.
65. J. H. Liu and T. Moriguchi, "Changes in free polyamines and gene expression during peach flower development," Biologia Plantarum, vol. 51, no. 3, pp. 530–532, 2007.
66. J. Wang, P. P. Sun, C. L. Chen, Y. Wang, X. Z. Fu, and J. H. Liu, "An arginine decarboxylase genePtADC from Poncirus trifoliata confers abiotic stress tolerance and promotes primary root growth inArabidopsis," Journal of Experimental Botany, vol. 62, no. 8, pp. 2899–2914, 2011.
67. J. H. Liu, I. Nakajima, andT. Moriguchi, "Effects of salt and osmotic stresses on free polyamine content and expression of polyamine biosynthetic genes in Vitis vinifera," Biologia Plantarum, vol. 55, no. 2, pp. 340–344, 2011.
68. Z. Hossain, M. F. López-Climent,V. Arbona, R. M. Pérez-Clemente, and A. Gómez-Cadenas, "Modulation of the antioxidant system in citrus under waterlogging and subsequent drainage,"Journal of Plant Physiology, vol. 166, no. 13, pp. 1391–1404, 2009.
69. J. Amudha and G. Balasubramani, "Recent molecular advances to combat abiotic stress tolerance in crop plants," Biotechnology and Molecular Biology Review, vol. 6, no. 2, pp. 31–58, 2011.
70. L. Tang, S. Y. Kwon, S. H. Kim et al., "Enhanced tolerance of transgenic potato plants expressing both superoxide dismutase and ascorbate peroxidase in chloroplasts against oxidative stress and high temperature," Plant Cell Reports, vol. 25, no. 12, pp. 1380–1386, 2006.

71. J. S. Rohila, R. K. Jain, and R. Wu, "Genetic improvement of Basmati rice for salt and drought tolerance by regulated expression of a barley Hva1 cDNA," Plant Science, vol. 163, no. 3, pp. 525–532, 2002.
72. Y. Y. Wu, Q. J. Chen, M. Chen, J. Chen, and X. C. Wang, "Salt-tolerant transgenic perennial ryegrass (Lolium perenne L.) obtained by Agrobacterium tumefaciens-mediated transformation of the vacuolar Na^+/H^+ antiporter gene," Plant Science, vol. 169, no. 1, pp. 65–73, 2005.
73. N. Sui, M. Li, S. J. Zhao, F. Li, H. Liang, and Q. W. Meng, "Overexpression of glycerol-3-phosphate acyltransferase gene improves chilling tolerance in tomato," Planta, vol. 226, no. 5, pp. 1097–1108, 2007.
74. G. Miller and R. Mittler, "Could heat shock transcription factors function as hydrogen peroxide sensors in plants?" Annals of Botany, vol. 98, no. 2, pp. 279–288, 2006.
75. P. Bhatnagar-Mathur, V. Vadez, and K. K. Sharma, "Transgenic approaches for abiotic stress tolerance in plants: retrospect and prospects," Plant Cell Reports, vol. 27, no. 3, pp. 411–424, 2008.
76. Y. Li, O. P. Dhankher, L. Carreira et al., "Overexpression of phytochelatin synthase in Arabidopsisleads to enhanced arsenic tolerance and cadmium hypersensitivity," Plant and Cell Physiology, vol. 45, no. 12, pp. 1787–1797, 2004.
77. R. Chaffai and H. Koyama, "Heavy metal tolerance in Arabidopsis thaliana," Advances in Botanical Research, vol. 6, pp. 1–49, 2011.
78. R. Munns, "Genes and salt tolerance: bringing them together," New Phytologist, vol. 167, no. 3, pp. 645–663, 2005.
79. K. Yamaguchi-Shinozaki and K. Shinozaki, "Transcriptional regulatory networks in cellular responses and tolerance to dehydration and cold stresses," Annual Review of Plant Biology, vol. 57, pp. 781–803, 2006.
80. T. Tzfira and V. Citovsky, "Agrobacterium-mediated genetic transformation of plants: biology and biotechnology," Current Opinion in Biotechnology, vol. 17, no. 2, pp. 147–154, 2006.
81. R. M. Pérez-Clemente and A. Gómez-Cadenas, "In vitro tissue culture, a tool for the study and breeding of plants subjected

to abiòtic stress conditions," Recent Advances in Plant In Vitro Culture, pp. 91–108, 2012.

82. L. E. Trujillo, M. Sotolongo, C. Menéndez et al., "SodERF3, a novel sugarcane ethylene responsive factor (ERF), enhances salt and drought tolerance when overexpressed in tobacco plants," Plant and Cell Physiology, vol. 49, no. 4, pp. 512–525, 2008. · ·
83. D. Fu, B. Huang, Y. Xiao, S. Muthukrishnan, and G. H. Liang, "Overexpression of barley hva1 gene in creeping bentgrass for improving drought tolerance," Plant Cell Reports, vol. 26, no. 4, pp. 467–477, 2007.
84. Q. C. Yan, M. S. Kuo, S. Li et al., "AGPAT6 is a novel microsomal glycerol-3-phosphate acyltransferase," Journal of Biological Chemistry, vol. 283, no. 15, pp. 10048–10057, 2008. · ·
85. J. Catinot, A. Buchala, E. Abou-Mansour, and J. P. Métraux, "Salicylic acid production in response to biotic and abiotic stress depends on isochorismate in Nicotiana benthamiana," FEBS Letters, vol. 582, no. 4, pp. 473–478, 2008.
86. X. Y. Yang, W. J. Jiang, and H. J. Yu, "The expression profiling of the lipoxygenase (LOX) family genes guring fruit development, abiotic stress and hormonal treatments in cucumber (Cucumis sativus L.)," International Journal of Molecular Sciences, vol. 13, pp. 2481–2500, 2012.
87. A. Andreou and I. Feussner, "Lipoxygenases—structure and reaction mechanism," Phytochemistry, vol. 70, no. 13-14, pp. 1504–1510, 2009.
88. J. A. Christianson, E. S. Dennis, D. J. Llewellyn, and I. W. Wilson, "ATAF NAC transcription factors: regulators of plant stress signaling," Plant Signaling and Behavior, vol. 5, no. 4, pp. 428–432, 2010.
89. M. Mazid, T. A. Khan, and F. Mohammad, "Role of secondary metabolites in defense mechanisms of plants," Biology and Medicine, vol. 3, no. 2, pp. 232–249, 2011.
90. A. Edreva, V. Velikova, T. Tsonev et al., "Stress-protective role of secondary metabolites: diversity of functions and mechanisms," Genetics and Plant Physiology, vol. 34, no. 1, pp. 67–78, 2008.

Chapter 9

Metabolic Engineering of Yeast and Plants for the Production of the Biologically Active Hydroxystilbene, Resveratrol

Philippe Jeandet,[1] Bertrand Delaunois,[1,2] Aziz Aziz,[2] David Donnez,[1,2] Yann Vasserot,[1] Sylvain Cordelier,[2] and Eric Courot[2]

[1]Laboratory of Enology and Applied Chemistry, Research Unit "Vines and Wines of Champagne," UPRES EA 4707, Faculty of Sciences, University of Reims, P.O. Box 1039, 51687 Reims Cedex 02, France
[2]Laboratory Stress, Defenses and Plant Reproduction, Research Unit "Vines and Wines of Champagne," UPRES EA 4707, Faculty of Sciences, University of Reims, P.O. Box 1039, 51687 Reims Cedex 02, France

ABSTRACT

Resveratrol, a stilbenic compound deriving from the phenyalanine / polymalonate route, being stilbene synthase the last and key enzyme of this pathway, recently has become the focus of a number of studies in medicine and plant physiology. Increased demand for this molecule for nutraceutical, cosmetic and possibly pharmaceutic uses, makes its production a necessity. In this context, the use of biotechnology through recombinant microorganisms and plants is particularly promising. Interesting results can indeed arise from the potential of genetically modified microorganisms as an alternative mechanism for producing resveratrol. Strategies used to tailoring yeast as they do not possess the genes that encode for the resveratrol pathway, will be described. On the other hand, most interest has centered in recent years, on STS gene transfer experiments from various origins to the genome of numerous plants. This work also presents a comprehensive review on plant molecular engineering with the STS gene, resulting in disease resistance against microorganisms and the enhancement of the antioxidant activities of several fruits in transgenic lines.

INTRODUCTION

Hydroxystilbenes (hereafter referred to as stilbenes) are natural phenolic compounds occurring in a number of plant families including Vitaceae, Dipterocarpaceae, Gnetaceae, Pinaceae, Poaceae, Fabaceae, Leguminoseae, and Cyperaceae [1]. Although polyphenolic compounds display an enormous chemical diversity, stilbenes seem to constitute a rather restricted group of molecules, the skeleton of which is based on resveratrol especially in Vitaceae and Fabaceae (Figure 1) [2]. Resveratrol is one of the most extensively studied natural products, doubtless for its large spectrum of biological activities in human health as a cardioprotective, an antitumor, a neuroprotective, and an antioxidant agent. Some of the resveratrol's properties have been associated with the benefits of a moderate consumption of red wine. Many roles have also been ascribed to resveratrol and its derivatives (Figure 1) in plants; namely, they constitute antimicrobial, deterrent or repellent compounds acting as allelochemicals or phytoalexins, protecting them from attacks by fungi, bacteria, nematodes, or herbivores [3–5].

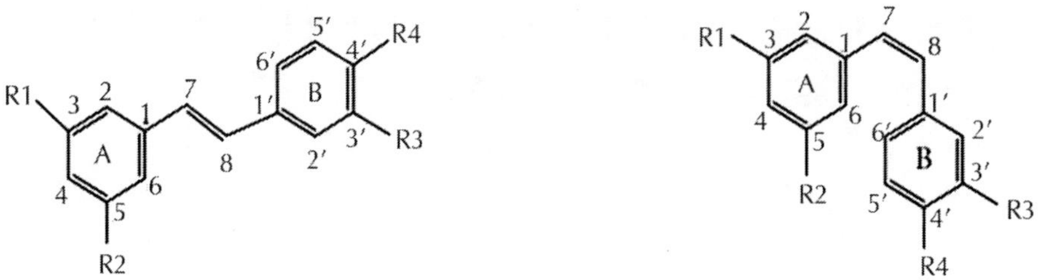

	R1	R2	R3	R4
1	OH	OH	H	OH
2	OH	GlcO	H	OH

	R1	R2	R3	R4
3	OH	OH	H	OH
4	OH	GlcO	H	OH

Figure 1. Molecular structures of trans- and cis-resveratrol 1 and 3, and trans- and cis-piceid, 2 and 4, respectively (GlcO = β-D-glucose).

According to these potential activities in plants and humans, the interest for resveratrol has increased. Currently, rising demand for resveratrol and derivatives for nutraceutical, cosmetic, and putatively pharmaceutical uses makes their production a necessity. Metabolic engineering for resveratrol production thus has significant commercial value. As all the genes encoding enzymes responsible for resveratrol biosynthesis have been cloned and characterized in detail, this makes molecular engineering of this compound relatively straightforward.

Most interest has now centered upon metabolic engineering of resveratrol in plants with the objectives of increasing tolerance of the latter to pathogenic microorganisms and improving the nutritional quality of food products through the expression of pharmaceutically active compounds in plants incapable of synthesizing resveratrol.

Microorganisms can also be used to heterologously express the resveratrol biosynthetic pathway to obtain this compound in valuable amounts. In this paper, we will discuss the potential of tailored microorganisms, specifically yeast, and plants for resveratrol production.

RESVERATROL IN HEALTH AND DISEASE

Resveratrol was identified in 1940 as a constitutive compound of the roots of white hellebore (Veratrum grandiflorum) [6]. Resveratrol is also present in high amounts in the roots of Polygonum cuspidatum used in traditional Chinese and Japanese medicine, and this compound was acclaimed for its wondrous effects for the treatment of human fungal diseases (suppurative dermatitis, gonorrhea favus, and athlete's foot), hyperlipidemia, atherosclerosis, or inflammations. Moreover, resveratrol recently has been shown to be a potent therapeutic agent [7–10].

First, there is considerable evidence that resveratrol acts both as a free radical scavenger [11, 12] and a potent antioxidant doubtless for its ability to promote the activities of numerous antioxidative enzymes [12,13]. Resveratrol inhibits lipid peroxidation which is an indicator of possible free radical damage to cellular membranes [14, 15]. Resveratrol may operate in a number of antioxidant mechanisms leading to the development of atherosclerosis, protecting, for example, LDL molecules against peroxidation [16–18] or reactive oxygen species production by blood platelets [19]. There is however some evidence that resveratrol exhibits pro-oxidant activity under certain experimental conditions (in the presence of copper ions) [20], causing oxidative DNA damage that may lead to cell cycle arrest or apoptosis [21].

Resveratrol is also considered as an antiproliferative agent for cancer [22], exerting an antitumor activity either as a cytostatic or as a cytotoxic agent in various types of cancer. Resveratrol induces cell death through positive induction of death receptor-mediated apoptosis [23–25], mitchondria-mediated apotosis [26], and nuclear (transcription) factor-mediated apotosis [27, 28].

Refractory disease and poor prognosis in many tumors have been related with high expression of antiapoptotic molecules such as survivin; resveratrol was proven to act at this level through survivin depletion [29]. Similarly, numerous works showed that resveratrol is a potent inhibitor of cell cycle progression, causing G1 phase arrest [30], inhibiting G0-G1 transition in human lymphocyte [31] or perturbing progression through S and G2 phases in cultured bovine artery endothelial cell proliferation [32]. Since the pioneering work of the group of Pezzuto [33], affording for the very first time evidence for the cancer chemoprotective activity of resveratrol on HL-60 and Hepa 1clc7 cells, many other models of human cancers were used, confirming these findings. The most frequently described mode of action for resveratrol concerns apoptosis (see above), the response depending on the expression of the tumor suppressor gene p53. A number of related factors can be modulated by resveratrol, such as activation of caspases, decreases in the antiapoptotic proteins Bcl-2 and Bcl-x^L [34, 35], increases in the proapoptotic proteins Bax [36], inhibition of cyclins and cyclin-dependent kinases, that is, proteins implicated in cell cycle progression [32,37], and interference with nuclear transcription factor kappa B (NF-κB)—and activator protein 1 (AP-1)—mediated cascades.

Most importantly, there are several works concerning both the antitumor effects and the antitumor mechanisms of resveratrol in vivo. In several studies, it was shown, namely, that resveratrol inhibits the development of skin cancer in vivo by topical applications, causing a significant reduction of the tumor diameter and the tumor incidence [33, 38, 39]. Moreover, resveratrol significantly inhibits the UV-B-mediated increase in skin thickness and skin edema [40]. On the other hand, resveratrol was proven not to be very effective in inhibiting the progression of leukemia in vivo, even when high doses were used [41], despite its antileukemic activity in vitro and though resveratrol can be converted to the known antileukemic related stilbene, piceatannol [42]. In regard to breast cancer, resveratrol was shown to reduce N-methyl-N-nitrosourea-induced mammary tumorigenesis in female rats at high doses (100 mg/kg body weight) though being uneffective at lower doses (10 mg/kg body weight) [43]. Otherwise, some authors have shown that resveratrol glucosides, namely, piceid (a 3-O-β-D-resveratrol glucoside), administered orally or intraperitoneally to mice, reduced tumor volume, tumor weight, and metastasis in lung carcinoma [44, 45]. A few investigations about the antitumor effects of resveratrol and derivatives against liver cancers have been reported to date, showing that this compound administered orally or intraperitoneally to rats caused a 25% reduction in tumor cell numbers and restrained hepatoma cell invasion but not proliferation [46, 47]. Finally, orally administration of resveratrol to mice was shown to prevent colonic tumor formation and reduce small intestinal tumors by 70% [48]. Neurologic benefits of resveratrol described experimentally concern the following diseases: cerebral ischemia in rats [49, 50], amyotrophic lateral sclerosis [51], Parkinson's disease in rats [52], spinal cord lesions in rabbits [53], brain edema and tumors in human cells [54], seizure in rats [55–57], pain and cognitive impairment in rats [58, 59]. According to Doré [60], the neuroprotective effects of resveratrol could be mediated by regulation of the heme oxygenase antioxidant systems in neurons, especially in case of age-related vascular dementia and Alzheimer's disease [61].

There are several mechanisms to support resveratrol as a cardioprotective agent such as inhibition of LDL peroxidation, reduction of the degree of neointimal hyperplasia and restenosis, inhibition of platelet aggregation together with chemoprevention of atherosclerogenesis. Resveratrol was proven to inhibit LDL peroxidation via antioxidant and free scavenging activities in ex vivo rat heart studies [62] and circumstantial evidence of its potent role in preventing atherosclerogenesis could thus be given [63, 64]. Resveratrol was shown to inhibit neointimal hyperplasia in several animal studies and vascular smooth muscle cell proliferation in a rabbit model of restenosis [65, 66] and to block platelet aggregation from high-cholesterol-fed rabbits [67].

Resveratrol exerts some activities against a range of bacteria affecting humans (Chlamydia pneumonia, the cause of human acute respiratory tract infections [68], and Helicobacter pylori, the main agent responsible for chronic gastritis and peptic ulcer disease [69, 70]). Resveratrol also inhibited the growth of several bacteria known to be major agents of human skin infections such as Staphylococcus aureus, Enterococcus faecalis, and Pseudomonas aeruginosa [71]. Similarly, resveratrol completely inhibits the growth of Neisseria gonorrhoeae (responsible for the sexually transmitted disease gonorrhoea) [72]. Stilbenes generally also have biocidal activities against plant fungal pathogens (namely, Botrytis cinerea, the causal organism for gray mold, Pyricularia oryzae, the agent of pyriculariosis, Plasmopara viticola, the causal agent for downy mildew [73–77], fungi associated with esca of grapevine [78]), and human fungal pathogens such as Candida albicans, an agent of candidiasis [79]. Finally, resveratrol has been shown to extend the lifespan of lower organisms, yeast [80], and metazoans [81], via the sirtuin/Sir2 family. More recently, a study has confirmed the conservation of these preventive and protective mechanisms controlling lifespan extension in higher organisms such as mice [82]. Using a similar approach, other works reached the same conclusions with regard to the protective effects of resveratrol against diet-induced obesity and insulin resistance [7, 82].

Otherwise, DNA-damaging products can induce premature senescence in cancer cells, limiting tumor development. However, senescent cancer cells may reenter the cell cycle and lead to tumor relapse. Recently, resveratrol was remarked by its ability to induce DNA damage in cancer cells. In fact, resveratrol suppressed viability and induced DNA damage in human head and neck squamous carcinoma cells [83]. Similarly, human squamous cancer cells treated with resveratrol were shown to express oxidative stress-mediated DNA damage [84].

BIOSYNTHESIS OF RESVERATROL

Stilbene phytoalexins, as flavonoid-type phytoalexins, are formed on the phenylalanine/polymalonate route, being the last step of this biosynthesis pathway catalyzed by stilbene synthase (STS) (Figure 2). Trans-resveratrol can be synthesized either starting with phenylalanine or from tyrosine, both pathways giving rise to para-coumaric acid through, respectively, the phenylalanine ammonia lyase (PAL) and the cinnamate 4 hydroxylase (C4H), or directly through the tyrosine ammonia lyase (TAL). A para-coumaric-acid coenzyme A ligase (4CL) transforms the para-coumaric acid into para-coumaroyl-CoA, which then leads to trans-resveratrol after condensation with three molecules of malonyl-CoA through STS activity [85]. STS belongs to the so-called type III of the polyketide synthase enzyme superfamily, a class of enzymes which carry out iterative condensation reactions with malonyl-CoA. [86, 87]. Resveratrol synthesis is induced in plants as a response to fungal infection [73, 74, 88], abiotic stresses (UV irradiation, metallic salts, methyl jasmonate) [89–93] as well as to natural compounds eliciting plant defense responses [94, 95] or nonpathogenic rhizobacteria [96, 97].

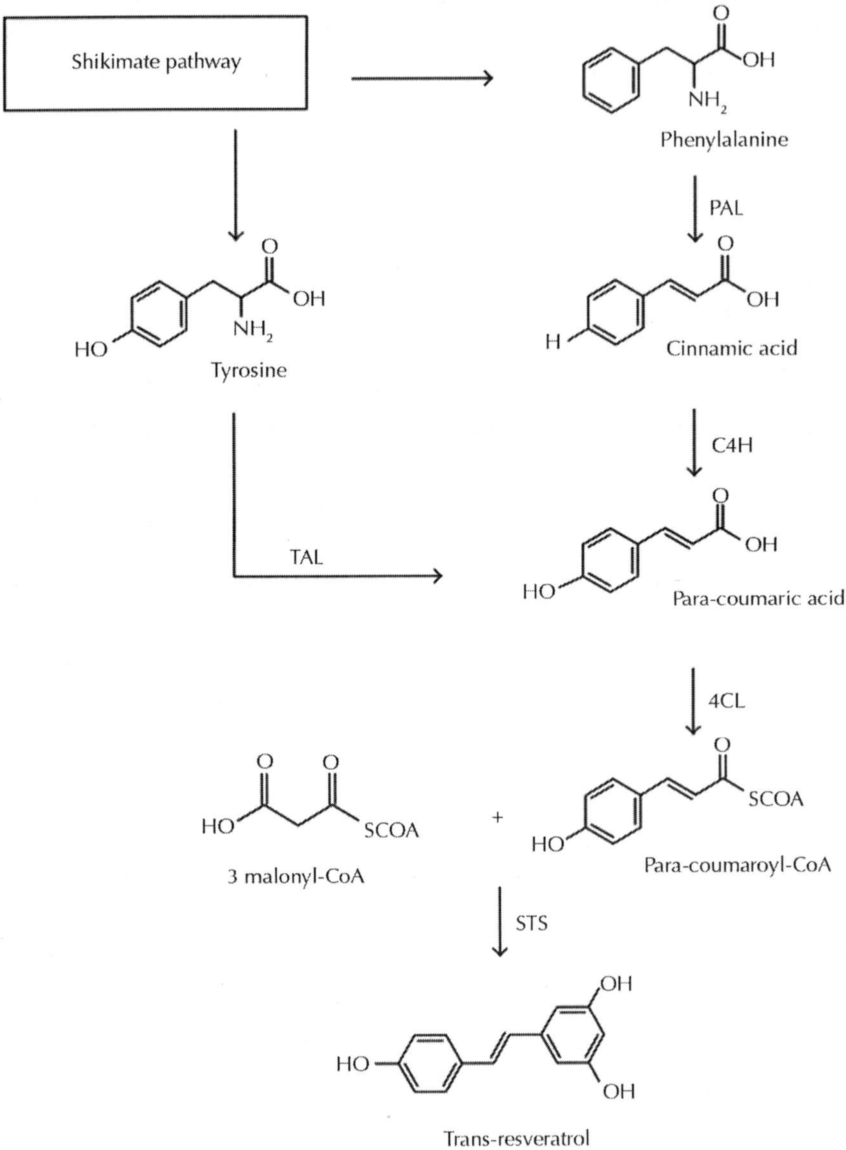

Figure 2: Biosynthesis of resveratrol via the phenylalanine/polymalonate pathway. PAL, phenylalanine ammonia lyase; TAL, tyrosine ammonia lyase; C4H, cinnamate-4-hydroxylase; 4CL, para-coumaric acid: coenzyme A ligase; STS, stilbene synthase.

STS is encoded by a multigene family mainly comprising the resveratrol-forming STS genes from grapevine (pSV21, pSV25, pSV696, pSV368, and StSy) [98], (Vst1, Vst2, Vst3) [99], the AhRS gene from

Arachis hypogea [100], and an STS-encoding gene from Parthenocissus henryana [101]. Other STS genes have also been isolated from pine [102–104], together with a stilbene synthase gene from Vitis riparia cv Gloire de Montpellier [105]. The Sb STS1 gene isolated from Sorghum is at present the only one STS gene described in a Monocotyledonous plant [106]. Knowledge of the resveratrol biosynthetic route thus paves the way for metabolic engineering in microorganisms or plants.

TAILORING YEAST FOR RESVERATROL BIOSYNTHESIS

Microorganisms are widely used biological systems whose engineering may be useful for the production of numerous valuable molecules and can lead to interesting applications for example in the wine industry [107–109]. Engineering bacteria or yeast for resveratrol might thus represent a valuable means of its production in large quantities. Their tailoring is necessary because microorganisms do not possess the genes that encode for the resveratrol pathway. In this paper, we will focus only on yeast tailoring for resveratrol production. Two main strategies have already been used to that end: (i) introducing the entire biosynthetic pathway using aromatic amino acids as substrates (L-phenylalanine or L-tyrosine) [110–113] and (ii) introducing specific genes, such as para-coumaroyl-CoA ligase and stilbene synthase, starting with para-coumaric acid as a substrate (Figure 2) [114–120].

Engineering the Entire Pathway

To obtain resveratrol from its precursors L-phenylalanine or L-tyrosine on the phenylpropanoid route appears to be the most promising option in terms of production cost. The entire resveratrol pathway has been introduced successfully into the oleaginous yeast Yarrowia lipolytica (ATCC 20362 strain) with the genes encoding for phenylalanine/tyrosine ammonia lyase (PAL/TAL), cinnamate-4-hydroxylase (4CH),para-coumaroyl-CoA ligase (4CL), and stilbene synthase (STS) activities [110] (Table 1). Metabolic engineering was completed by constitutive expression of malonyl CoA, a precursor for both naringenin-chalcone, the first C15 intermediate in the flavonoid route, and stilbene synthesis.

Genes that encode for each required enzyme from different plant species (grapevine, peanuts) have been tested. In the best performing yeast, genes that encode for the PAL and TAL from Rhodotorula glutinis, the 4CL fromStreptomyces coelicolor, and an STS from Vitis sp. have been used, and the resulting production of resveratrol reached in this system 1.46 mg/L [110].

Table 1. Metabolic engineering of resveratrol in yeast

Microorganisms/ species	Introduced gene(s)	Origin of genes	Resveratrol quantity	References
Yeast Yarrowia lipolytica	PAL/TAL, C4H, 4CL,STS	Rhodotorula glutinis (PAL/TAL), Streptomyces coelicolor (4CL), Vitis sp. (STS)	1.46 mg/L	[110]
Yeast Saccharomyces cerevisiae	PAL, C4H, 4CL, STS	Arabidopsis thaliana (PAL, C4H, 4CL), Rheum tataricum (STS)	not detectable	[111]
Yeast Saccharomyces cerevisiae	TAL, 4CL::STS fusion protein	Rhodobacter sphaeroides (TAL), Arabidopsis thaliana (4CL), Vitis vinifera (STS)	5.25 mg/L	[116]
Yeast Saccharomyces cerevisiae	4CL, STS	Populus trichocarpa× Populus deltoides (4CL),Vitis vinifera (STS)	1.45 mg/L	[114]
Yeast Saccharomyces cerevisiae	4CL, STS	Nicotiana tabacum (4CL), Vitis vinifera (STS)	5.8 mg/L	[115]
Yeast Saccharomyces cerevisiae	4CL1, STS	Arabidopsis thaliana (4CL), Vitis vinifera (STS)	262–391 mg/L	[117]

Yeast Saccharomyces cerevisiae with phenylalanine	PAL, CPR, C4H, 4CL, STS	Populus trichocarpa×P. deltoides (PAL, CPR[a])Glycine max (C4H, 4CL), Vitis vinifera (STS)	0.29 mg/L	[113]
Yeast Saccharomyces cerevisiae with para-coumaric acid	PAL, CPR, C4H, 4CL, STS	Populus trichocarpa×P. deltoides (PAL, CPR)Glycine max (C4H, 4CL), Vitis vinifera (STS)	0.31 mg/L	[113]
Yeast Saccharomyces cerevisiae	TAL, 4CL::STS fusion protein, araE transporter	Rhodobacter sphaeroides (TAL), Escherichia coli(araE), Arabidopsis thaliana (4CL), Vitis vinifera (STS)	3.1 mg/L 1.27 mg/L (without the araE transporter)	[119]
Yeast Saccharomyces cerevisiae	TAL, 4CL::STS fusion protein	Rhodobacter sphaeroides (TAL), Arabidopsis thaliana (4CL), Vitis vinifera (STS)	14.4 mg/L	[120]

[a]CPR: Cytochrome P450 Reductase.

Current efforts to increase resveratrol bioproduction in yeast have focused, so far not surprisingly, on heterologous expression of the genes encoding for the enzymes involved in the phenylpropanoid pathway and for STS, in baker yeast Saccharomyces cerevisiae [111, 113–120]. Namely, the entire resveratrol pathway has been introduced in S. cerevisiae (CEN.PK113-5D strain), but also in molds such as Aspergillus niger(FGSC A913 strain) and A. oryzae (MG1363 strain). In these cases, pathway expression started with the PAL enzyme. The ability of tailored S. cerevisiae to produce resveratrol was characterized, but it was not synthesized in measurable amounts in this system. [111]. More recently, Trantas et al. [113] constructed the complete resveratrol biosynthetic pathway in S. cerevisiae to produce resveratrol from phenylalanine. When the medium was supplemented with 10 mM of phenylalanine, the strain produced 0.29 mg/L trans-resveratrol after about 120 h of cultivation. One can thus admit that the introduction of the complete resveratrol pathway in yeast leads to a low production of this compound when compared to some other data obtained in engineered bacteria where resveratrol synthesis can reach up to 100–170 mg/L [121, 122].

Introducing Selective Genes

An alternative strategy to engineering the entire pathway is directed towards transforming microorganisms with selective genes [114–120] (Table 1). The yeast strain S. cerevisiae FY23 was transformed with both the4CL and STS genes under utilization of para-coumaric acid as a precursor (added to the culture medium). However, in this system, resveratrol production remained low (1.45 mg/L) [114]. Interestingly, in another study, authors reported that transformation of S. cerevisiae (CEN.PK113-3b strain) with the 4CL gene from tobacco and the STS gene from grapevine enabled it to produce resveratrol in higher quantities (5.8 mg/L) [115]. A S. cerevisiae (WAT11 strain) cotransformed with 4CL and STS constructs and fed with para-coumaric acid only produced 0.65 mg/L resveratrol, but, when the 4CL and the STS-encoding genes that were added to the yeast genome were submitted to protein fusion, yeast expressing 4CL::STS fusion protein exhibited a 10-fold increase in resveratrol production (5.25 mg/L resveratrol) compared to the coexpression of 4CL and STS [116]. This underlines the importance of the spatial localization of these two related enzymes [116]. In another study of the same group [120], authors have utilized a synthetic scaffold to recruit the 4CL1 and STS enzymes of the resveratrol pathway to improve resveratrol production in S. cerevisiae. A 5-fold improvement of the resveratrol production was obtained over the nonscaffolded control, and a 2.7-fold increase (14.4 mg/L within 96 h incubation) was finally observed over the previous reported study with protein fusion [116]. This work clearly demonstrated that synthetic scaffolds can be used for the optimization of engineered metabolic pathway.

Most importantly, it has recently been shown that yeast cells expressing 4CL, STS genes together with thearaE gene encoding for a high-capacity Escherichia coli transporter, but with no affinity for resveratrol itself, could enhance resveratrol accumulation. Yeast cells carrying the araE gene produced up to 3.1 mg/L, that is, 2.44-fold higher resveratrol than the control cells [119]. Such an engineered yeast was also proven to increase the resveratrol content in a white wine during the fermentation process [119].

It should be noted that the efficacy of recombinant microorganisms for resveratrol production depends on various factors, such as the species and the strain, the origin of the transferred genes, culture conditions as well as other parameters such as plasmids or precursors used (Table 1). A recent study [117] has shown that fermenting yeast expressing the para-coumaroyl-coenzyme A ligase (4CL1) from Arabidopsis thaliana and the stilbene synthase from Vitis vinifera in a rich medium could considerably improve resveratrol production, rising from a few milligrams per liter to 262 mg/L in rich medium using a laboratory strain. Moreover, resveratrol amounts reached up to 391 mg/L when fermentation was achieved with an industrial Brazilian sugar cane yeast [117].

Taken together, these data indicate that the levels of resveratrol that can be produced by microorganisms remain low, although optimization of the processes might be possible.

TAILORING PLANTS FOR RESVERATROL BIOSYNTHESIS

As resveratrol is a potent phytoalexin against plant pathogens and can enhance plant resistance to microbial disease, earlier applications of resveratrol engineering focused primarily on this antimicrobial potential [123,124]. Tailoring plants for resveratrol synthesis thus constitutes the second aspect of this paper. Particular attention will be given in this section to the choice of the promoters and the enhancer elements used to improve STS transcriptional activity in the transgenes as well as the biological benefits of resveratrol production in terms of enhancement of antioxidant activity in fruits and legumes.

We have seen that tailoring yeast needs the introduction of the entire resveratrol pathway (requiring at least 4 or 5 genes) or the introduction of selective genes (requiring at least 2 genes) depending on the precursor used, phenylalanine or para-coumaric acid. Tailoring plants for resveratrol synthesis appear to be very simple, since STS is a key enzyme of resveratrol synthesis utilizing as substrates precursor molecules that are present throughout the plant kingdom. The introduction of a single gene is thus sufficient to synthesize resveratrol in heterologous plant species. A lot of transformations were then operated to investigate the potential of stilbene biosynthetic genes to confer resistance to pathogens or to increase their nutritional values [123, 124].

Production of Resveratrol in Transgenic Plants: Gene and Promoter Options

The first gene transfer experiment was performed by the group of Kindl with a complete STS gene fromArachis hypogea introduced into tobacco [100], leading to resveratrol accumulation in response to short-wavelength ultraviolet light. This experiment was then continued by the same group by the transfer of two grapevine STS genes, Vst1 and Vst2, in tobacco, conferring to the plant a higher resistance to Botrytis cinereainfection. This work constitutes the first report of a disease resistance resulting from foreign phytoalexin expression in a novel plant [125]. Since this pioneering work, STS genes have been transferred to a number of plants, including rice [126], barley and wheat [127–130], alfalfa [131], kiwifruit [132], grapevine [133,134] apple [135, 136], aspen [137], papaya [138], white poplar [139], oilseed rape [140], banana [141],Rehmannia [142] tomato [143–147], Arabidopsis [148], lettuce [101], pea [149], and hop [150] (Table 2).

Table 2. Metabolic engineering of stilbene synthase in plants, and resulting effects on stilbene levels, resistance to pathogens, and antioxidant activities

Plant/species	Introduced gene(s)	Promoter	Produced stilbene(s)	Stilbene quantity (mg/kg of FW)	Biological activity	References
Tobacco (Nicotiana tabacum L.)	Arachis hypogea STS	Stress-induced promoter	Resveratrol	—	—	[100]
	Grapevine Vst1 and Vst2	Stress responsive pVst1	Resveratrol	400	Resistance to Botrytis cinerea	[125]
	Chimeric STSgene	Constitutive CaMV 35S	Resveratrol	50 to 290	Altered flower morphology, male sterility	[152]
Rice (Oryza sativa L.)	Grapevine Vst1	Stress responsive pVst1	—	—	Resistance to Pyricularia oryzae ?	[126]
Wheat (Triticum aestivum L.)	Grapevine Vst1	Combination pVst1 +35S enhancer	—	—	Resistance to Botrytis cinerea	[127, 129]
	Chimeric STSgene	Maize ubiquitin promoter	Resveratrol	2	—	[128]
	Grapevine Vst1and Vst2	Combination pVst1 +35S enhancer	Unknown derivative stilbene compounds	35 to 190	Resistance to Puccinia recondita and Septoria nodorum	[130]

Barley (Hordeum vulgare L.)	Grapevine Vst1	Combination pVst1 +35S enhancer	—	—	Resistance to Botrytis cinerea	[127]
Alfalfa (Medicago sativa L.)	Arachis hypogeaSTS gene(AhRS)	Constitutive CaMV 35S	Piceid	0.5 to 20	Resistance to Phoma medicaginis	[131]
Arabidopsis thaliana L.	Sorghum SbSTS1	Constitutive CaMV 35S	Piceid	584	—	[106, 148]
Kiwi (Actinidia deliciosa)	pSV25	Constitutive CaMV 35S	Piceid	20 to 182	No resistance to Botrytis cinerea	[132]
Grapevine (Vitis vinifera L.)	Grapevine Vst1	Fungus inducible msPR 10.1	Resveratrol		In vitro resistance to Botrytis cinerea	[133]
	Vitis pseudoreticulata STS	Constitutive CaMV 35S	Resveratrol	2.586	Not determined	[134]
Apple (Malus domestica)	Grapevine Vst1	Stress responsive pVst1	Unknown resveratrol-glycoside	—	—	[135]
	Grapevine Vst1	Stress responsive pVst1	Piceid	3 to 7 for non-UV-irradiated fruit and 23 to 62 for UV-irradiated fruit	No influence on other phenolic compounds	[136]

Tomato (Lycopersicon esculentum Mill.)	Grapevine Vst1and Vst2	Stress responsive pVst1	Resveratrol	—	Resistance to Phytophthora infestans No resistance toBotrytis cinerea andAlternaria solani	[143]
	Grapevine StSy	Constitutive pCaMV 35S	Resveratrol and piceid	4 to 53	Antioxidant primary metabolism and increase in total antioxidant activity	[144]
	Grapevine StSy	Constitutive pCaMV 35S	Resveratrol and piceid	0.1 to 1.2	Enhancement of natural antiradical properties	[145]
	Grapevine StSy	Constitutive pCaMV 35S	Resveratrol and piceid	0.42 to 126 depending on the stage of ripening and fruit samples	Differences in rutin, naringenin, and chlorogenic acid contents	[146]
	Grapevine StSy	Fruit-specific promoter TomLoxB	Resveratrol and piceid		Increases in total antioxidant capability and ascorbic acid content	[147]

Rehmannia glutinosa Libosch.	Arachis hypogea AhRS3	Constitutive pCaMV 35S	Resveratrol and piceid	22 to 116 up to 650 with stress treatment	Antioxidant capabilities Resistance to Fusarium oxysporum	[142]
Lettuce (Lactuca sativa L.)	Parthenocissus henryanaSTS	Constitutive pCaMV 35S	Resveratrol	56.4	Effect on Hela cell morphology	[101]
Pea (Pisum. sativum L.)	Grapevine Vst1	Stress responsive pVst1	Occurrence of two resveratrol-glucoside compounds	0.53 to 5.2	—	[149]
White poplar (Populus alba L.)	Grapevine StSy	Constitutive pCaMV 35S	Piceid	309 to 615	No in vitro resistance toMelampsora pulcherrima	[137, 139]
Papaya (Carica papaya L.)	Grapevine Vst1	Stress responsive pVst1	Resveratrol glucoside	54	Resistance to Phytophthora palmivora	[138]
Oilseed rape (Brassica napus L.)	Grapevine Vst1	Tissue specific p-nap	Resveratrol glucoside	361 to 616	Food quality improvement: high piceid rate content and reduction of sinapate esters	[140]
Hop (Humulus lupulus L.)	Grapevine Vst1	Constitutive pCaMV 35S	Piceid, unknown stilbene astringin, resveratrol	490 to 560	Higher amounts of flavonoids and acids	[150]

In grapevine, genome sequencing has revealed a large array of STS genes, with 43 genes identified and 20 of these being shown to be expressed [151]. But to date, only a few STS genes from grapevine are used for the metabolic engineering of plants, being Vst1 and Stsy the most commonly genes chosen (Table 2). STS-encoding genes from other plants have also been used, notably the AhRS gene from Arachis hypogea [100,131], the SbSTS1 gene from Sorghum bicolor [106, 148], and an STS-encoding gene from Parthenocissus henryana [101]. Chimeric genes or a combination of two STS encoding genes, based on Vst1 and Vst2, can increase significantly resveratrol production in the transformed lines [128, 152].

On a practical point of view, the modulation of gene expression is mainly controlled by the promoter chosen to drive the transgene. By now, a limited number of promoters (the constitutive promoter pCaMV35S, its own stress-responsive promoter pVst1, the fungus-inducible promoter pPR10.1 or the tissue specific promoter p-nap) have been used for the expression of STS-encoding genes upon plant transformation (Table 2 and references therein). As expected, the pCaMV35S promoter, which is the most commonly used to overexpress a transgene in plants [152], triggered strong and constitutive stilbene accumulation (Table 2). In this case, stilbene synthesis is higher than that observed with inducible promoters, but, as a consequence, causes a drastic depletion of the endogenous pools of precursors. In grapevine, the pVst1 promoter which is induced either by biotic factors (pathogens, elicitors) or abiotic stresses (wounding, UV light), allows high stilbene production without interfering with secondary biosynthetic pathways. Generally, it appears thus preferable to transform plants with a construct having a pathogen-inducible promoter in order to avoid depletion of other pathways.

Expression of the STS gene may be optimized by the utilization of enhancer elements [127, 130] and/or heterologous promoters [131–135, 140, 152] to improve STS transcription. Combination of the pVst1 promoter with the 35S enhancer element [127, 130] or use of a chimeric promoter resulting, for example, from a fusion between the alfalfa pPR10.1 promoter and the pVst1 promoter can lead to higher expression of the transgene without affecting promoter inducibility or specific expression patterns [127], or to an increased production of resveratrol upon fungal infection (reaching 5–100-fold the levels found in nontransgenic leaves [133]). Finally, tissue-specific promoters such as the p-nap seed-specific napin promoter can be used to induce stilbene production [140].

At this stage of the discussion, it appears important to underline that the choice of the promoter for STSgene expression should be performed depending on the expected results: when the enhancement of plant resistance against pathogens through resveratrol production is searched, thus one can recommend the choice of a strong constitutive promoter or a pathogen-inducible promoter (see above). As far as may concern this review, when the improvement of food products is sought through, for example the increase of the antioxidant activity of the transformed plants, a tissue-specific [140, 147] or inducible promoter would be a better option.

STS genes were transferred to plants by means of Agrobacterium spp. in tomato [143], grapevine [133], kiwifruits [132] or particle bombardment in barley and wheat [127–129]. However, as epigenetic modifications may occur and lead to expression variability, decreasing the ability of the transgenes for stilbene synthesis, the selection of plants with a single gene insertion will be more appropriate. As a consequence, the use of Agrobacterium-mediated transformation, which leads to lower transgene insertion numbers, should preferentially be chosen [124].

Stilbene Production and Biological Benefits of Resveratrol Synthesis in STS Engineered Plants

Expression of STS genes resulted in resveratrol accumulation in transgenic lines. The obtained stilbene amounts are generally higher than those reached in engineered yeast, ranging from a few mg/kg fresh weight to hundreds of mg/kg fresh weight (see Table 2). Both free resveratrol and its glycosylated forms can be recovered in plant extracts [144]: piceid, a 3-O-β-D-resveratrol glucoside, occurred in different plant species transformed with STS genes (Table 2) [106, 132, 136, 139, 140, 142, 145, 146, 148, 150]. However, levels of accumulated stilbenes depend on the plant species (probably because of different endogenous pools of enzymes or precursors, as well as differences in secondary metabolism pathways), the promoter used (constitutive or inducible promoters), the ripening stage in case of transgenic fruits [136, 144] and the age of the organs [125, 132].

From a practical viewpoint, the ectopic production of resveratrol observed in transformed plants can lead to broad-spectrum resistance against fungi in transgenic plants, but disparate effects were observed (Table 2). Some works indeed reported transformations improving the resistance of rice to Pyricularia oryzae [126], tomato to Phytophthora infestans [143], barley and wheat to B. cinerea [127, 129], wheat to Oidium tuckeri[127, 129], alfalfa to Phoma medicaginis [131] and papaya to Phytophthora palmivora [138]. Conversely, in some others, no resistance was observed after transforming the plants with stilbene synthase genes. For example, transformation of white poplar (Populus alba) with STS, leading to the accumulation of piceid, does not confer any increased resistance to rust disease (Melaspora pulcherrima) [137, 139]. Similarly, no increased resistance against B. cinerea has been observed in STS transgenic kiwi plants [132].

With regard to the improvement of the nutritional value of agricultural crops and fruits, there are already several studies reporting an increase of the antioxidant activities in transgenic tomatoes and apples overexpressing STS-encoding genes [136, 144, 147]. In tomato, for example, resveratrol accumulation in transgenic fruit increased their global antioxidant activities, as well as their contents in other well-known antioxidants such as ascorbic acid and glutathione [144]. Antioxidant activity, as a consequence of resveratrol accumulation, was also shown to suffer a two-fold increase in transgenic tomato fruit versuscontrols [145]. Moreover, a correlation was found between resveratrol concentrations and antioxidant activities in ripe and unripe fruits [145]. More recently [147], tomato plants expressing a stilbene synthase gene (StSy) under the control of a fruit-specific promoter (promoter TomLoxB) were shown to accumulate resveratrol and piceid in the skin of the mature fruits, being the resveratrol content of the plants transformed with the specific promoter TomLoxB 20-fold lower than that of plants previously transformed with the constitutive pCaMV35S promoter [144]. However, both the total antioxidant capability and the ascorbic acid content were increased in the transformed fruits. These results explain the higher capability of transgenic fruits to counteract the proinflammatory effects of phorbol ester in monocyte–macrophages viathe inhibition of induced cyclo-oxygenase-2 enzyme [147]. This last example constitutes a nice illustration of what can be expected from molecular engineering of resveratrol in plants in terms of improvement of the nutritional value of fruits or food products.

CONCLUSIONS

Increased demand for resveratrol for nutraceutical, cosmetic and possibly pharmaceutic uses makes its production from sustainable sourcing a necessity. In this context, the use of biotechnology through recombinant microorganisms and plants is particularly promising [5, 112, 118]. Interesting results can indeed arise from the potential of genetically modified microorganisms as an alternative mechanism for producing resveratrol, as this compound can be synthesized directly in recombinant yeast (the subject of the present review) but also in bacteria, such as Escherichia coli [112, 118]. Use of recombinant bacteria or yeast is of interest for the food industry, which could produce resveratrol in large quantities in biofermentators. Tailoring yeast can also receive direct applications in winemaking, as for example, fermentation engineering to produce resveratrol in wine (or to increase the wine resveratrol content) [119]. Otherwise, a transgenic yeast expressing a gene for a glycosyl hydrolase capable of liberating free resveratrol from its glucoside form has been reported as well to increase resveratrol amounts in wine [153] or the wine-related antioxidant content [114]. Beside the fact that disease resistance can be obtained following expression of STS-encoding genes, molecular engineering of plants with resveratrol may also lead to food products comprising edible legumes, cereals, or fruits, which can be ingested with their potential clinical benefits, by humans. Taken together, these results suggest the overall relevance of metabolic engineering of resveratrol.

REFERENCES

1. M. Morales, A. Ros Barcelo, and M. A. Pedreno, "Plant stilbenes: recent advances in their chemistry and biology," Advances in Plant Physiology, vol. 3, pp. 39–70, 2000.
2. P. Jeandet, A. C. Breuil, M. Adrian et al., "HPLC analysis of

grapevine phytoalexins coupling photodiode array detection and fluorometry," Analytical Chemistry, vol. 69, pp. 5172–5177, 1997.

3. P. Jeandet, A. C. Douillet-Breuil, R. Bessis, S. Debord, M. Sbaghi, and M. Adrian, "Phytoalexins from the vitaceae: biosynthesis, phytoalexin gene expression in transgenic plants, antifungal activity, and metabolism," Journal of Agricultural and Food Chemistry, vol. 50, no. 10, pp. 2731–2741, 2002.
4. J. Chong, A. Poutaraud, and P. Hugueney, "Metabolism and roles of stilbenes in plants," Plant Science, vol. 177, no. 3, pp. 143–155, 2009.
5. P. Jeandet, B. Delaunois, A. Conreux et al., "Biosynthesis, metabolism, molecular engineering, and biological functions of stilbene phytoalexins in plants," BioFactors, vol. 36, no. 5, pp. 331–341, 2010.
6. M. J. Takaoka, "Of the phenolic substances of white hellebore (Veratrum grandiflorum Loes. Fil.),"Journal of the Faculty of Science Hokkaido Imperial University, vol. 3, pp. 1–16, 1940.
7. J. A. Baur and D. A. Sinclair, "Therapeutic potential of resveratrol: the in vivo evidence," Nature Reviews Drug Discovery, vol. 5, no. 6, pp. 493–506, 2006.
8. R. E. King, J. A. Bomser, and D. B. Min, "Bioactivity of resveratrol," Comprehensive Reviews in Food Science and Food Safety, vol. 5, no. 3, pp. 65–70, 2006.
9. P. Saiko, A. Szakmary, W. Jaeger, and T. Szekeres, "Resveratrol and its analogs: defense against cancer, coronary disease and neurodegenerative maladies or just a fad?" Mutation Research, vol. 658, no. 1-2, pp. 68–94, 2008.
10. J. M. Pezzuto, "The phenomenon of resveratrol: redefining the virtues of promiscuity," Annals of the New York Academy of Sciences, vol. 1215, no. 1, pp. 123–130, 2011.
11. J. Martinez and J. J. Moreno, "Effect of resveratrol, a natural polyphenolic compound, on reactive oxygen species and prostaglandin production," Biochemical Pharmacology, vol. 59, no. 7, pp. 865–870, 2000.
12. G. A. Losa, "Resveratrol modulates apoptosis and oxidation in human blood mononuclear cells,"European Journal of Clinical Investigation, vol. 33, no. 9, pp. 818–823, 2003.

13. M. Jang and J. M. Pezzuto, "Cancer chemopreventive activity of resveratrol," Drugs under Experimental and Clinical Research, vol. 25, no. 2-3, pp. 65–77, 1999.
14. Y. J. Cai, J. G. Fang, L. P. Ma, L. Yang, and Z. L. Liu, "Inhibition of free radical-induced peroxidation of rat liver microsomes by resveratrol and its analogues," Biochimica et Biophysica Acta, vol. 1637, no. 1, pp. 31–38, 2003.
15. S. S. Leonard, C. Xia, B. H. Jiang et al., "Resveratrol scavenges reactive oxygen species and effects radical-induced cellular responses," Biochemical and Biophysical Research Communications, vol. 309, no. 4, pp. 1017–1026, 2003.
16. E. N. Frankel, A. L. Waterhouse, and J. E. Kinsella, "Inhibition of human LDL oxidation by resveratrol," The Lancet, vol. 341, no. 8852, pp. 1103–1104, 1993.
17. L. Belguendouz, L. Fremont, and A. Linard, "Resveratrol inhibits metal ion-dependent and independent peroxidation of porcine low-density lipoproteins," Biochemical Pharmacology, vol. 53, no. 9, pp. 1347–1355, 1997.
18. F. Ferroni, A. Maccaglia, D. Pietraforte, L. Turco, and M. Minetti, "Phenolic antioxidants and the protection of low density lipoprotein from peroxynitrite-mediated oxidations at physiologic CO_2,"Journal of Agricultural and Food Chemistry, vol. 52, no. 10, pp. 2866–2874, 2004.
19. B. Olas, H. M. Zbikowska, B. Wachowicz, T. Krajewski, A. Buczyński, and A. Magnuszewska, "Inhibitory effect of resveratrol on free radical generation in blood platelets," Acta Biochimica Polonica, vol. 46, no. 4, pp. 961–966, 1999.
20. A. Ahmad, F. A. Syed, S. Singh, and S. M. Hadi, "Prooxidant activity of resveratrol in the presence of copper ions: mutagenicity in plasmid DNA," Toxicology Letters, vol. 159, no. 1, pp. 1–12, 2005.
21. M. Athar, J. H. Back, L. Kopelovich, D. R. Bickers, and A. L. Kim, "Multiple molecular targets of resveratrol: anti-carcinogenic mechanisms," Archives of Biochemistry and Biophysics, vol. 486, no. 2, pp. 95–102, 2009.
22. M. Athar, J. H. Back, X. Tang et al., "Resveratrol: a review of preclinical studies for human cancer prevention," Toxicology and Applied Pharmacology, vol. 224, no. 3, pp. 274–283, 2007.

23. M. V. Clément, J. L. Hirpara, S. H. Chawdhury, and S. Pervaiz, "Chemopreventive agent resveratrol, a natural product derived from grapes, triggers CD95 signaling-dependent apoptosis in human tumor cells," Blood, vol. 92, no. 3, pp. 996–1002, 1998.
24. D. Delmas, C. Rébé, O. Micheau et al., "Redistribution of CD95, DR4 and DR5 in rafts accounts for the synergistic toxicity of resveratrol and death receptor ligands in colon carcinoma cells," Oncogene, vol. 23, no. 55, pp. 8979–8986, 2004.
25. J. L. Su, M. T. Lin, C. C. Hong et al., "Resveratrol induces FasL-related apotosis through Cdc42 activation of ASK1/JNK-dependent signaling pathway in human leukemia HL-60 cells,"Carcinogenesis, vol. 25, pp. 2005–2013, 2004.
26. J. Dörrie, H. Gerauer, Y. Wachter, and S. J. Zunino, "Resveratrol induces extensive apoptosis by depolarizing mitochondrial membranes and activating caspase-9 in acute lymphoblastic leukemia cells," Cancer Research, vol. 61, no. 12, pp. 4731–4739, 2001.
27. N. Kuwajerwala, E. Cifuentes, S. Gautam, M. Menon, E. R. Barrack, and G. P. Veer Reddy, "Resveratrol induces prostate cancer cell entry into S phase and inhibits DNA synthesis," Cancer Research, vol. 62, no. 9, pp. 2488–2492, 2002.
28. E. Pozo-Guisado, A. Alvarez-Barrientos, S. Mulero-Navarro, B. Santiago-Josefat, and P. M. Fernandez-Salguero, "The antiproliferative activity of resveratrol results in apoptosis in MCF-7 but not in MDA-MB-231 human breast cancer cells: cell-specific alteration of the cell cycle," Biochemical Pharmacology, vol. 64, no. 9, pp. 1375–1386, 2002.
29. S. Fulda and K. M. Debatin, "Sensitization for tumor necrosis factor-related apoptosis-inducing ligand-induced apoptosis by the chemopreventive agent resveratrol," Cancer Research, vol. 64, no. 1, pp. 337–346, 2004.
30. Z. Estrov, S. Shishodia, S. Faderl et al., "Resveratrol blocks interleukin-1β-induced activation of the nuclear transcription factor NF-κB, inhibits proliferation, causes S-phase arrest, and induces apoptosis of acute myeloid leukemia cells," Blood, vol. 102, no. 3, pp. 987–995, 2003.
31. T. C. Hsieh, D. Halicka, X. Lu et al., "Effects of resveratrol on the G0-G1 transition and cell cycle progression of mitogenically

stimulated human lymphocytes," Biochemical and Biophysical Research Communications, vol. 297, no. 5, pp. 1311–1317, 2002.

32. T. C. Hsieh, G. Juan, Z. Darzynkiewicz, and J. M. Wu, "Resveratrol increases nitric oxide synthase, induces accumulation of p53 and p21WAF1/CIPI1, and suppresses cultured bovine pulmonary artery endothelial cell proliferation by perturbing progression through S and G2," Cancer Research, vol. 59, no. 11, pp. 2596–2601, 1999.
33. M. Jang, L. Cai, G. O. Udeani et al., "Cancer chemopreventive activity of resveratrol, a natural product derived from grapes," Science, vol. 275, no. 5297, pp. 218–220, 1997.
34. J. W. Park, Y. J. Choi, S. I. Suh et al., "Bcl-2 overexpression attenuates resveratrol-induced apoptosis in U937 cells by inhibition of caspase-3 activity," Carcinogenesis, vol. 22, no. 10, pp. 1633–1639, 2001.
35. H. B. Zhou, Y. Yan, Y. N. Sun, and J. R. Zhu, "Resveratrol induces apoptosis in human esophageal carcinoma cells," World Journal of Gastroenterology, vol. 9, no. 3, pp. 408–411, 2003.
36. Y. A. Kim, W. H. Lee, T. H. Choi, S. H. Rhee, K. Y. Park, and Y. H. Choi, "Involvement of p21WAF1/CIPI1, pRB, Bax and NF-kappaB in induction of growth arrest and apoptosis by resveratrol in human lung carcinoma A549 cells," International Journal of Oncology, vol. 23, no. 4, pp. 1143–1149, 2003.
37. N. Ahmad, V. M. Adhami, F. Afaq, D. K. Feyes, and H. Mukhtar, "Resveratrol causes WAF-1/p21-mediated G1-phase arrest of cell cycle and induction of apoptosis in human epidermoid carcinoma A431 cells," Clinical Cancer Research, vol. 7, no. 5, pp. 1466–1473, 2001.
38. G. J. Kapadia, M. A. Azuine, H. Tokuda et al., "Chemopreventive effect of resveratrol, sesamol, sesame oil and sunflower oil in the Epstein-Barr virus early antigen activation assay and the mouse skin two-stage carcinogenesis," Pharmacological Research, vol. 45, no. 6, pp. 499–505, 2002.
39. S. Pervaiz, "Resveratrol- from the bottle to the bedside?" Leukemia and Lymphoma, vol. 40, no. 5-6, pp. 491–498, 2001.

40. F. Afaq, V. M. Adhami, and N. Ahmad, "Prevention of short-term ultraviolet B radiation-mediated damages by resveratrol in SKH-1 hairless mice," Toxicology and Applied Pharmacology, vol. 186, no. 1, pp. 28–37, 2003.

41. X. Gao, Y. X. Xu, G. Divine, N. Janakiraman, R. A. Chapman, and S. C. Gautam, "Disparate in vitroand in vivo antileukemic effects of resveratrol, a natural polyphenolic compound found in grapes,"Journal of Nutrition, vol. 132, no. 7, pp. 2076–2081, 2002.

42. G. A. Potter, L. H. Patterson, E. Wanogho et al., "The cancer preventative agent resveratrol is converted to the anticancer agent piceatannol by the cytochrome P450 enzyme CYP1b1," British Journal of Cancer, vol. 86, no. 5, pp. 774–778, 2002.

43. K. P. L. Bhat, D. Lantvit, K. Christov, R. G. Mehta, R. C. Moon, and J. M. Pezzuto, "Estrogenic and antiestrogenic properties of resveratrol in mammary tumor models," Cancer Research, vol. 61, no. 20, pp. 7456–7463, 2001.

44. Y. Kimura and H. Okuda, "Effects of naturally occurring stilbene glucosides from medicinal plants and wine, on tumour growth and lung metastasis in Lewis lung carcinoma-bearing mice," Journal of Pharmacy and Pharmacology, vol. 52, no. 10, pp. 1287–1295, 2000.

45. Y. Kimura and H. Okuda, "Resveratrol isolated from Polygonum cuspidatum root prevents tumor growth and metastasis to lung and tumor-induced neovascularization in lewis lung carcinoma-bearing mice," Journal of Nutrition, vol. 131, no. 6, pp. 1844–1849, 2001.

46. N. Carbo, P. Costelli, F. M. Baccino, F.J. Lopez-Soriano, and J. M. Argiles, "Resveratrol, a natural product present in wine, decreases tumour growth in a rat tumour model," Biochemical and Biophysical Research Communications, vol. 254, pp. 739–743, 1999.

47. Y. Kozuki, Y. Miura, and K. Yagasaki, "Resveratrol suppresses hepatoma cell invasion independently of its anti-proliferative action," Cancer Letters, vol. 167, no. 2, pp. 151–156, 2001.

48. Y. Schneider, B. Duranton, F. Gossé, R. Schleiffer, N. Seiler, and F. Raul, "Resveratrol inhibits intestinal tumorigenesis and

modulates host-defense-related gene expression in an animal model of human familial adenomatous polyposis," Nutrition and Cancer, vol. 39, no. 1, pp. 102–107, 2001.

49. S. S. Huang, M. C. Tsai, C. L. Chih, L. M. Hung, and S. K. Tsai, "Resveratrol reduction of infarct size in Long-Evans rats subjected to focal cerebral ischemia," Life Sciences, vol. 69, no. 9, pp. 1057–1065, 2001.
50. K. Sinha, G. Chaudhary, and Y. Kumar Gupta, "Protective effect of resveratrol against oxidative stress in middle cerebral artery occlusion model of stroke in rats," Life Sciences, vol. 71, no. 6, pp. 655–665, 2002.
51. S. N. Wu, "Large-conductance Ca^{2+}-activated K^+ channels: physiological role and pharmacology,"Current Medicinal Chemistry, vol. 10, no. 8, pp. 649–661, 2003.
52. J. Karlsson, M. Emgård, P. Brundin, and M. J. Burkitt, "trans-Resveratrol protects embryonic mesencephalic cells from tert-butyl hydroperoxide: electron paramagnetic resonance spin trapping evidence for a radical scavenging mechanism," Journal of Neurochemistry, vol. 75, no. 1, pp. 141–150, 2000.
53. U. Kiziltepe, N. N. D. Turan, U. Han, A. T. Ulus, and F. Akar, "Resveratrol, a red wine polyphenol, protects spinal cord from ischemia-reperfusion injury," Journal of Vascular Surgery, vol. 40, no. 1, pp. 138–145, 2004.
54. G. Nicolini, R. Rigolio, A. Scuteri et al., "Effect of trans-resveratrol on signal transduction pathways involved in paclitaxel-induced apoptosis in human neuroblastoma SH-SY5Y cells," Neurochemistry International, vol. 42, no. 5, pp. 419–429, 2003.
55. M. Virgili and A. Contestabile, "Partial neuroprotection of in vivo excitotoxic brain damage by chronic administration of the red wine antioxidant agent, trans-resveratrol in rats," Neuroscience Letters, vol. 281, no. 2-3, pp. 123–126, 2000.
56. Y. K. Gupta, G. Chaudhary, K. Sinha, and A. K. Srivastava, "Protective effect of resveratrol against intracortical FeCl3-induced model of posttraumatic seizures in rats," Methods and Findings in Experimental and Clinical Pharmacology, vol. 23, no. 5, pp. 241–244, 2001.

57. Y. K. Gupta, G. Chaudhary, and A. K. Srivastava, "Protective effect of resveratrol against pentylenetetrazole-induced seizures and its modulation by an adenosinergic system," Pharmacology, vol. 65, no. 3, pp. 170–174, 2002.
58. M. Gentilli, J. X. Mazoit, H. Bouaziz et al., "Resveratrol decreases hyperalgesia induced by carrageenan in the rat hind paw," Life Sciences, vol. 68, no. 11, pp. 1317–1321, 2001.
59. M. Sharma and Y. K. Gupta, "Chronic treatment with trans resveratrol prevents intracerebroventricular streptozotocin induced cognitive impairment and oxidative stress in rats,"Life Sciences, vol. 71, no. 21, pp. 2489–2498, 2002.
60. S. Doré, "Decreased activity of the antioxidant heme oxygenase enzyme: implications in ischemia and in Alzheimer's disease," Free Radical Biology and Medicine, vol. 32, no. 12, pp. 1276–1282, 2002.
61. T. S. Anekonda, "Resveratrol-a boon for treating Alzheimer's disease?" Brain Research Reviews, vol. 52, no. 2, pp. 316–326, 2006.
62. L. M. Hung, M. J. Su, W. K. Chu, C. W. Chiao, W. F. Chan, and J. K. Chen, "The protective effect of resveratrols on ischaemia-reperfusion injuries of rat hearts is correlated with antioxidant efficacy,"British Journal of Pharmacology, vol. 135, no. 7, pp. 1627–1633, 2002.
63. T. Hayek, B. Fuhrman, J. Vaya et al., "Reduced progression of atherosclerosis in apolipoprotein E-deficient mice following consumption of red wine, or its polyphenols quercetin or catechin, is associated with reduced susceptibility of LDL to oxidation and aggregation," Arteriosclerosis, Thrombosis, and Vascular Biology, vol. 17, no. 11, pp. 2744–2752, 1997.
64. J. A. Vinson, K. Teufel, and N. Wu, "Red wine, dealcoholized red wine, and especially grape juice, inhibit atherosclerosis in a hamster model," Atherosclerosis, vol. 156, no. 1, pp. 67–72, 2001.
65. J. Zou, Y. Huang, K. Cao et al., "Effect of resveratrol on intimal hyperplasia after endothelial denudation in an experimental rabbit model," Life Sciences, vol. 68, no. 2, pp. 153–163, 2000.

66. Z. H. Mnjoyan and K. Fujise, "Profound negative regulatory effects by resveratrol on vascular smooth muscle cells: a role of p53-p21WAF1/CIPI1 pathway," Biochemical and Biophysical Research Communications, vol. 311, no. 2, pp. 546–552, 2003.
67. Z. Wang, Y. Huang, J. Zou, K. Cao, Y. Xu, and J. M. Wu, "Effects of red wine and wine polyphenol resveratrol on platelet aggregation in vivo and in vitro," International Journal of Molecular Medicine, vol. 9, no. 1, pp. 77–79, 2002.
68. C. Schriever, S. L. Pendland, and G. B. Mahady, "Red wine, resveratrol, Chlamydia pneumoniae and the French connection," Atherosclerosis, vol. 171, no. 2, pp. 379–380, 2003.
69. G. B. Mahady and S. L. Pendland, "Resveratrol inhibits the growth of Helicobacter pyloriin vitro,"American Journal of Gastroenterology, vol. 95, no. 7, p. 1849, 2000.
70. G. B. Mahady, S. L. Pendland, and L. R. Chadwick, "Resveratrol and red wine extracts inhibit the growth of CagA+ strains of Helicobacter pylori in vitro," American Journal of Gastroenterology, vol. 98, no. 6, pp. 1440–1441, 2003.
71. M. M. Y. Chan, "Antimicrobial effect of resveratrol on dermatophytes and bacterial pathogens of the skin," Biochemical Pharmacology, vol. 63, no. 2, pp. 99–104, 2002.
72. J. J. Docherty, M. M. Fu, and M. Tsai, "Resveratrol selectively inhibits Neisseria gonorrhoeae andNeisseria meningitidis," Journal of Antimicrobial Chemotherapy, vol. 47, no. 2, pp. 243–244, 2001.
73. P. Langcake and R. J. Pryce, "The production of resveratrol by Vitis vinifera and other members of the Vitaceae as a response to infection or injury," Physiological Plant Pathology, vol. 9, no. 1, pp. 77–86, 1976.
74. P. Langcake, "Disease resistance of Vitis spp. and the production of the stress metabolites resveratrol,ε-viniferin, α-viniferin and pterostilbene," Physiological Plant Pathology, vol. 18, pp. 213–226, 1981.
75. M. Adrian, P. Jeandet, J. Veneau, L. A. Weston, and R. Bessis, "Biological activity of resveratrol, a stilbenic compound from grapevines, against Botrytis cinerea, the causal agent for gray mold," Journal of Chemical Ecology, vol. 23, no. 7, pp. 1689–1702, 1997.

76. M. Adrian, H. Rajaei, P. Jeandet, J. Veneau, and R. Bessis, "Resveratrol oxidation in Botrytis cinerea conidia," Phytopathology, vol. 88, no. 5, pp. 472–4776, 1998.
77. M. Adrian and P. Jeandet, "Resveratrol as an antifungal agent," in Resveratrol in Health and Disease, B. B. Aggarwal and S. Shishodia, Eds., pp. 475–497, CRC Press, Boca Raton, Fla, USA, 2006.
78. M. Alessandro, S. Di Marco, F. Osti, and A. Cesari, "Bioassays on the activity of resveratrol, pterostilbene and phosphorous acid towards fungi associated with esca of grapevine," Phytopathologia Mediterranea, vol. 39, no. 3, pp. 357–365, 2000.
79. H. J. Jung, I. A. Hwang, W. S. Sung, et al., "Fungicidal effect of resveratrol on human infectious fungi," Archives of Pharmacal Research, vol. 28, no. 5, pp. 557–560, 2005.
80. K. T. Howitz, K. J. Bitterman, H. Y. Cohen et al., "Small molecule activators of sirtuins extendSaccharomyces cerevisiae lifespan," Nature, vol. 425, no. 6954, pp. 191–196, 2003.
81. J. G. Wood, B. Rogina, S. Lavu et al., "Sirtuin activators mimic caloric restriction and delay ageing in metazoans," Nature, vol. 430, no. 7000, pp. 686–689, 2004.
82. J. L. Barger, T. Kayo, J. M. Vann et al., "A low dose of dietary resveratrol partially mimics caloric restriction and retards aging parameters in mice," PLoS ONE, vol. 3, no. 6, Article ID e2264, 2008.
83. A. Tyagi, M. Gu, T. Takahata et al., "Resveratrol selectively induces DNA damage, independent of Smad 4 expression, in its efficacy against human head and neck squamous cell carcinoma," Clinical Cancer Research, vol. 17, pp. 5402–5411, 2011.
84. J. H. Back, H. R. Rezvani, Y. Zhu et al., "Cancer cell survival following DNA damage-mediated premature senescence is regulated by mammalian target of rapamycin (mTOR)-dependent inhibition of sirtuin," Journal of Biological Chemistry, vol. 286, no. 21, pp. 19100–19108, 2011.
85. P. Langcake and R. J. Pryce, "The production of resveratrol and the viniferins by grapevines in response to ultraviolet irradiation," Phytochemistry, vol. 16, no. 8, pp. 1193–1196, 1977.

86. M. B. Austin and J. P. Noel, "The chalcone synthase superfamily of type III polyketide synthases,"Natural Product Reports, vol. 20, no. 1, pp. 79–110, 2003.
87. O. Yu and J. M. Jez, "Nature's assembly line: biosynthesis of simple phenylpropanoids and polyketides," Plant Journal, vol. 54, no. 4, pp. 750–762, 2008.
88. P. Jeandet, R. Bessis, M. Sbaghi, and P. Meunier, "Production of the phytoalexin resveratrol by grapes as a response to Botrytis attack under natural conditions," Journal of Phytopathology, vol. 143, pp. 135–139, 1995.
89. M. Adrian, P. Jeandet, R. Bessis, and J. M. Joubert, "Induction of phytoalexin (resveratrol) synthesis in grapevine leaves treated with aluminum chloride ($AlCl_3$)," Journal of Agricultural and Food Chemistry, vol. 44, no. 8, pp. 1979–1981, 1996.
90. A. C. Douillet-Breuil, P. Jeandet, M. Adrian, and R. Bessis, "Changes in the phytoalexin content of various Vitis spp. in response to ultraviolet C elicitation," Journal of Agricultural and Food Chemistry, vol. 47, no. 10, pp. 4456–4461, 1999.
91. M. Adrian, P. Jeandet, A. C. Douillet-Breuil, L. Tesson, and R. Bessis, "Stilbene content of matureVitis vinifera berries in response to UV-C elicitation," Journal of Agricultural and Food Chemistry, vol. 48, no. 12, pp. 6103–6105, 2000.
92. R. Schubert, R. Fischer, R. Hain et al., "An ozone-responsive region of the grapevine resveratrol synthase promoter differs from the basal pathogen-responsive sequence," Plant Molecular Biology, vol. 34, no. 3, pp. 417–426, 1997.
93. S. Krisa, F. Larronde, H. Budzinski, A. Decendit, G. Deffieux, and J. M. Mérillon, "Stilbene production by Vitis vinifera cell suspension cultures: methyl jasmonate induction and ^{13}C biolabeling," Journal of Natural Products, vol. 62, no. 12, pp. 1688–1690, 1999.
94. A. Aziz, B. Poinssot, X. Daire et al., "Laminarin elicits defense responses in grapevine and induces protection against Botrytis cinerea and Plasmopara viticola," Molecular Plant-Microbe Interactions, vol. 16, no. 12, pp. 1118–1128, 2003.
95. A. Aziz, P. Trotel-Aziz, L. Dhuicq, P. Jeandet, M. Couderchet, and G. Vernet, "Chitosan oligomers and copper sulfate induce

grapevine defense reactions and resistance to gray mold and downy mildew," Phytopathology, vol. 96, no. 11, pp. 1188–1194, 2006.

96. B. W. M. Verhagen, P. Trotel-Aziz, M. Couderchet, M. Höfte, and A. Aziz, "Pseudomonas spp.-induced systemic resistance to Botrytis cinerea is associated with induction and priming of defence responses in grapevine," Journal of Experimental Botany, vol. 61, no. 1, pp. 249–260, 2010.

97. B. Verhagen, P. Trotel-Aziz, P. Jeandet, F. Baillieul, and A. Aziz, "Improved resistance against Botrytis cinerea by grapevine-associated bacteria that induce a prime oxidative burst and phytoalexin production," Phytopathology, vol. 101, no. 7, pp. 768–777, 2011.

98. F. Melchior and H. Kindl, "Coordinate- and elicitor-dependent expression of stilbene synthase and phenylalanine ammonia-lyase genes in Vitis cv. Optima," Archives of Biochemistry and Biophysics, vol. 288, no. 2, pp. 552–557, 1991.

99. W. Wiese, B. Vornam, E. Krause, and H. Kindl, "Structural organization and differential expression of three stilbene synthase genes located on a 13 kb grapevine DNA fragment," Plant Molecular Biology, vol. 26, no. 2, pp. 667–677, 1994.

100. R. Hain, B. Bieseler, H. Kindl, G. Schröder, and R. Stöcker, "Expression of a stilbene synthase gene inNicotiana tabacum results in synthesis of the phytoalexin resveratrol," Plant Molecular Biology, vol. 15, no. 2, pp. 325–335, 1990.

101. S. Liu, Y. Hu, X. Wang, J. Zhong, and Z. Lin, "High content of resveratrol in lettuce transformed with a stilbene synthase gene of Parthenocissus henryana," Journal of Agricultural and Food Chemistry, vol. 54, no. 21, pp. 8082–8085, 2006.

102. R. Preisig-Müller, A. Schwekendiek, I. Brehm, H. J. Reif, and H. Kindl, "Characterization of a pine multigene family containing elicitor-responsive stilbene synthase genes," Plant Molecular Biology, vol. 39, no. 2, pp. 221–229, 1999.

103. A. Kodan, H. Kuroda, and F. Sakai, "Simultaneous expression of stilbene synthase genes in Japanese red pine (Pinus densiflora) seedlings," Journal of Wood Science, vol. 47, no. 1, pp. 58–62, 2001.

104. A. Kodan, H. Kuroda, and F. Sakai, "A stilbene synthase from Japanese red pine (Pinus densiflora): implications for phytoalexin accumulation and down-regulation of flavonoid biosynthesis,"Proceedings of the National Academy of Sciences of the United States of America, vol. 99, no. 5, pp. 3335–3339, 2002.
105. P. H. Goodwin, T. Hsiang, and L. Erickson, "A comparison of stilbene and chalcone synthases including a new stilbene synthase gene from Vitis riparia cv. Gloire de Montpellier," Plant Science, vol. 151, no. 1, pp. 1–8, 2000.
106. C. K. Y. Yu, K. Springob, J. Schmidt et al., "A stilbene synthase gene (SbSTS1) is involved in host and nonhost defense responses in sorghum," Plant Physiology, vol. 138, no. 1, pp. 393–401, 2005.
107. I. S. Pretorius, "Tailoring wine yeast for the new millennium: novel approaches to the ancient art of winemaking," Yeast, vol. 16, pp. 675–729, 2000.
108. J. A. Chemler and M. A. Koffas, "Metabolic engineering for plant natural product biosynthesis in microbes," Current Opinion in Biotechnology, vol. 19, no. 6, pp. 597–605, 2008.
109. Y. Wang, S. Chen, and O. Yu, "Metabolic engineering of flavonoids in plants and microorganisms,"Applied Microbiology and Biotechnology, vol. 91, pp. 949–956, 2011.
110. L. L. Huang, Z. Xue, and Q. Q. Zhu, "Method for the production of resveratrol in a recombinant oleaginous microorganism," World Patent WO 2006125000 A2, 2006.
111. M. Katz, J. Förster, H. David, et al., "Metabolically engineered cells for the production of resveratrol or an oligomeric or glycosidically-bound derivative thereof," US Patent US 2008/0286844A1, 2008.
112. D. Donnez, P. Jeandet, C. Clément, and E. Courot, "Bioproduction of resveratrol and stilbene derivatives by plant cells and microorganisms," Trends in Biotechnology, vol. 27, no. 12, pp. 706–713, 2009.
113. E. Trantas, N. Panopoulos, and F. Ververidis, "Metabolic engineering of the complete pathway leading to heterologous biosynthesis of various flavonoids and stilbenoids in Saccharomyces cerevisiae," Metabolic Engineering, vol. 11, no. 6, pp. 355–366, 2009.

114. J. V. W. Becker, G. O. Armstrong, M. J. Van Der Merwe, M. G. Lambrechts, M. A. Vivier, and I. S. Pretorius, "Metabolic engineering of Saccharomyces cerevisiae for the synthesis of the wine-related antioxidant resveratrol," FEMS Yeast Research, vol. 4, no. 1, pp. 79–85, 2003.
115. J. Beekwilder, R. Wolswinkel, H. Jonker, R. Hall, C. H. De Rie Vos, and A. Bovy, "Production of resveratrol in recombinant microorganisms," Applied and Environmental Microbiology, vol. 72, no. 8, pp. 5670–5672, 2006.
116. Y. Zhang, S. Z. Li, J. Li et al., "Using unnatural protein fusions to engineer resveratrol biosynthesis in yeast and mammalian cells," Journal of the American Chemical Society, vol. 128, no. 40, pp. 13030–13031, 2006.
117. T. Sydor, S. Schaffer, and E. Boles, "Considerable increase in resveratrol production by recombinant industrial yeast strains with use of rich medium," Applied and Environmental Microbiology, vol. 76, no. 10, pp. 3361–3363, 2010.
118. Y. Wang, H. Chen, and O. Yu, "Metabolic engineering of resveratrol and other longevity boosting compounds," BioFactors, vol. 36, no. 5, pp. 394–400, 2010.
119. Y. Wang, C. Halls, J. Zhang, M. Matsuno, Y. Zhang, and O. Yu, "Stepwise increase of resveratrol biosynthesis in yeast Saccharomyces cerevisiae by metabolic engineering," Metabolic Engineering, vol. 13, pp. 455–463, 2011.
120. Y. Wang and O. Yu, "Synthetic scaffolds increased resveratrol biosynthesis in engineered yeast cells,"Journal of Biotechnology, vol. 157, pp. 258–260, 2012.
121. K. T. Watts, P. C. Lee, and C. Schmidt-Dannert, "Biosynthesis of plant-specific stilbene polyketides in metabolically engineered Escherichia coli," BMC Biotechnology, vol. 6, article 22, 2006.
122. Y. Katsuyama, N. Funa, I. Miyahisa, and S. Horinouchi, "Synthesis of unnatural flavonoids and stilbenes by exploiting the plant biosynthetic pathway in Escherichia coli," Chemistry and Biology, vol. 14, no. 6, pp. 613–621, 2007.
123. C. Halls and O. Yu, "Potential for metabolic engineering of resveratrol biosynthesis," Trends in Biotechnology, vol. 26, no. 2, pp. 77–81, 2008.

124. B. Delaunois, S. Cordelier, A. Conreux, C. Clément, and P. Jeandet, "Molecular engineering of resveratrol in plants," Plant Biotechnology Journal, vol. 7, no. 1, pp. 2–12, 2009.
125. R. Hain, H. J. Reif, E. Krause et al., "Disease resistance results from foreign phytoalexin expression in a novel plant," Nature, vol. 361, no. 6408, pp. 153–156, 1993.
126. P. Stark-Lorenzen, B. Nelke, G. Hänßler, H. P. Mühlbach, and J. E. Thomzik, "Transfer of a grapevine stilbene synthase gene to rice (Oryza sativa L.)," Plant Cell Reports, vol. 16, no. 10, pp. 668–673, 1997.
127. G. Leckband and H. Lörz, "Transformation and expression of a stilbene synthase gene of Vitis viniferaL. in barley and wheat for increased fungal resistance," Theoretical and Applied Genetics, vol. 96, no. 8, pp. 1004–1012, 1998.
128. S. Fettig and D. Hess, "Expression of a chimeric stilbene synthase gene in transgenic wheat lines,"Transgenic Research, vol. 8, no. 3, pp. 179–189, 1999.
129. H. Liang, J. Zheng, X. Duan et al., "A transgenic wheat with a stilbene synthase gene resistant to powdery mildew obtained by biolistic method," Chinese Science Bulletin, vol. 45, no. 7, pp. 634–638, 2000.
130. L. Serazetdinova, K. H. Oldach, and H. Lörz, "Expression of transgenic stilbene synthases in wheat causes the accumulation of unknown stilbene derivatives with antifungal activity," Journal of Plant Physiology, vol. 162, no. 9, pp. 985–1002, 2005.
131. J. D. Hipskind and N. L. Paiva, "Constitutive accumulation of a resveratrol-glucoside in transgenic alfalfa increases resistance to Phoma medicaginis," Molecular Plant-Microbe Interactions, vol. 13, no. 5, pp. 551–562, 2000.
132. S. Kobayashi, C. K. Ding, Y. Nakamura, I. Nakajima, and R. Matsumoto, "Kiwifruits (Actinidia deliciosa) transformed with a Vitis stilbene synthase gene produce piceid (resveratrol-glucoside),"Plant Cell Reports, vol. 19, no. 9, pp. 904–910, 2000.
133. P. Coutos-Thévenot, B. Poinssot, A. Bonomelli et al., "In vitro tolerance to Botrytis cinerea of grapevine 41B rootstock in transgenic plants expressing the stilbene synthase Vst 1 gene under the control of a pathogen-inducible PR 10 promoter,"

Journal of Experimental Botany, vol. 52, no. 358, pp. 901–910, 2001.

134. C. Fan, N. Pu, X. Wang et al., "Agrobacterium-mediated genetic transformation of grapevine (Vitis vinifera L.) with a novel stilbene synthase gene from Chinese wild Vitis pseudoreticulata," Plant Cell, Tissue and Organ Culture, vol. 92, no. 2, pp. 197–206, 2008.

135. I. Szankowski, K. Briviba, J. Fleschhut, J. Schönherr, H. J. Jacobsen, and H. Kiesecker, "Transformation of apple (Malus domestica Borkh.) with the stilbene synthase gene from grapevine (Vitis vinifera L.) and a PGIP gene from kiwi (Actinidia deliciosa)," Plant Cell Reports, vol. 22, no. 2, pp. 141–149, 2003.

136. S. Rühmann, D. Treutter, S. Fritsche, K. Briviba, and I. Szankowski, "Piceid (resveratrol glucoside) synthesis in stilbene synthase transgenic apple fruit," Journal of Agricultural and Food Chemistry, vol. 54, no. 13, pp. 4633–4640, 2006.

137. S. K. Seppänen, L. Syrjälä, K. Von Weissenberg, T. H. Teeri, L. Paajanen, and A. Pappinen, "Antifungal activity of stilbenes in in vitro bioassays and in transgenic Populus expressing a gene encoding pinosylvin synthase," Plant Cell Reports, vol. 22, no. 8, pp. 584–593, 2004.

138. Y. J. Zhu, R. Agbayani, M. C. Jazckson, C. S. Tang, and P. H. Moore, "Expression of the grapevine stilbene synthase gene VST 1 in papaya provides increased resistance against diseaes caused byPhytophthora palmivora," Planta, vol. 12, pp. 807–812, 2004.

139. A. Giorcelli, F. Sparvoli, F. Mattivi et al., "Expression of the stilbene synthase (StSy) gene from grapevine in transgenic white poplar results in high accumulation of the antioxidant resveratrol glucosides," Transgenic Research, vol. 13, no. 3, pp. 203–214, 2004.

140. A. Hüsken, A. Baumert, C. Milkowski, H. C. Becker, D. Strack, and C. Möllers, "Resveratrol glucoside (Piceid) synthesis in seeds of transgenic oilseed rape (Brassica napus L.)," Theoretical and Applied Genetics, vol. 111, no. 8, pp. 1553–1562, 2005.

141. J. Vishnevetsky, M. Flaishman, Y. Cohen, et al., "Transgenic disease resistant banana," World Patent, WO2005/047515, 2005.

142. J. D. Lim, S. J. Yun, I. M. Chung, and C. Y. Yu, "Resveratrol synthase transgene expression and accumulation of resveratrol glycoside in Rehmannia glutinosa," Molecular Breeding, vol. 16, no. 3, pp. 219–233, 2005.
143. J. E. Thomzik, K. Stenzel, R. Stöcker, P. H. Schreier, R. Hain, and D. J. Stahl, "Synthesis of a grapevine phytoalexin in transgenic tomatoes (Lycopersicon esculentum Mill.) conditions resistance againstPhytophthora infestans," Physiological and Molecular Plant Pathology, vol. 51, no. 4, pp. 265–278, 1997.
144. G. Giovinazzo, L. D'Amico, A. Paradiso, R. Bollini, F. Sparvoli, and L. DeGara, "Antioxidant metabolite profiles in tomato fruit constitutively expressing the grapevine stilbene synthase gene,"Plant Biotechnology Journal, vol. 3, no. 1, pp. 57–69, 2005.
145. R. Morelli, S. Das, A. Bertelli et al., "The introduction of the stilbene synthase gene enhances the natural antiradical activity of Lycopersicon esculentum mill," Molecular and Cellular Biochemistry, vol. 282, no. 1-2, pp. 65–73, 2006.
146. I. Nicoletti, A. De Rossi, G. Giovinazzo, and D. Corradini, "Identification and quantification of stilbenes in fruits of transgenic tomato plants (Lycopersicon esculentum Mill.) by reversed phase HPLC with photodiode array and mass spectrometry detection," Journal of Agricultural and Food Chemistry, vol. 55, no. 9, pp. 3304–3311, 2007.
147. A. D'Introno, A. Paradiso, E. Scoditti et al., "Antioxidant and anti-inflammatory properties of tomato fruits synthesizing different amounts of stilbenes," Plant Biotechnology Journal, vol. 7, no. 5, pp. 422–429, 2009.
148. C. K. Y. Yu, C. N. W. Lam, K. Springob, J. Schmidt, I. K. Chu, and C. Lo, "Constitutive accumulation of cis-piceid in transgenic Arabidopsis overexpressing a sorghum stilbene synthase gene," Plant and Cell Physiology, vol. 47, no. 7, pp. 1017–1021, 2006.
149. A. Richter, H. J. Jacobsen, A. De Kathen et al., "Transgenic peas (Pisum sativum) expressing polygalacturonase inhibiting protein from raspberry (Rubus idaeus) and stilbene synthase from grape (Vitis vinifera)," Plant Cell Reports, vol. 25, no. 11, pp. 1166–1173, 2006.

150. A. Schwekendiek, O. Spring, A. Heyerick et al., "Constitutive expression of a grapevine stilbene synthase gene in transgenic hop (Humulus lupulus L.) yields resveratrol and its derivatives in substantial quantities," Journal of Agricultural and Food Chemistry, vol. 55, no. 17, pp. 7002–7009, 2007.
151. O. Jaillon, J. M. Aury, B. Noel et al., "The grapevine genome sequence suggests ancestral hexaploidization in major angiosperm phyla," Nature, vol. 449, no. 7161, pp. 463–467, 2007.
152. R. Fischer, I. Budde, and R. Hain, "Stilbene synthase gene expression causes changes in flower colour and male sterility in tobacco," Plant Journal, vol. 11, no. 3, pp. 489–498, 1997.
153. L. Gonzalez-Candelas, J. V. Gil, R. M. Lamuela-Raventos, and D. Ramon, "The use of transgenic yeasts expressing a gene encoding a glycosyl-hydrolase as a tool to increase resveratrol content in wine," International Journal of Food Microbiology, vol. 59, pp. 179–183, 2000.

Chapter 10

Recent Advances in Medicago truncatula Genomics

Jean-Michel Ané,[1] Hongyan Zhu,[2] and Julia Frugoli[3]

[1]Department of Agronomy, University of Wisconsin, Madison, WI 53706, USA

[2]Department of Plant and Soil Sciences, University of Kentucky, Lexington, KY 40546, USA

[3]Department of Genetics and Biochemistry, Clemson University, 100 Jordan Hall, Clemson, SC 29634, USA

ABSTRACT

Legume rotation has allowed a consistent increase in crop yield and consequently in human population since the antiquity. Legumes will also be instrumental in our ability to maintain the sustainability of our agriculture while facing the challenges of increasing food and biofuel demand. Medicago truncatula and Lotus japonicus have emerged during the last decade as two major model systems for legume biology. Initially developed to dissect plant-microbe symbiotic interactions and especially legume nodulation, these two models are now widely used in a variety of biological fields from plant physiology and development to population genetics and structural genomics. This review highlights the genetic and genomic tools available to the M. truncatula community. Comparative genomic approaches to transfer biological information between model systems and legume crops are also discussed.

INTRODUCTION

Legumes are usually defined by their typical flower structure and the ability of many of them to form root nodules in presence of symbiotic bacteria named rhizobia. With more than 18 000 species, legumes are found from the artic circle to the tropics and include many crops of agronomic importance for grain production, pasture, and forestry [1, 2]. The ability of more than 88% of legumes to obtain nitrogen from the air through root nodules was probably a major determinant in this evolutionary, ecological, and economical success [3]. Interestingly, the study of symbiotic associations with rhizobia as well as with arbuscular mycorrhizal (AM) fungi also drove the development of two model legumes: Medicago truncatula Gaertner and Lotus japonicus (Regel) K. Larsen.

While M. truncatula is an annual medic from the Trifolieae tribe and a close relative of alfalfa and clovers, L. japonicus belongs to the Loteae and is more distant from cultivated cool season legumes than M. truncatula. This phylogenetic distance to economically important crops is critical in the choice of M. truncatula by many researchers and support by numerous funding agencies. The use of both model legumes allows unique comparative genomic studies within the legume family as well as the comparison between two patterns of root nodule development: indeterminate with a persistent nodule meristem in the case of M. truncatula and determinate in L. japonicus. Unfortunately, these two models belong to the same cool season legumes (Galegoid clade), whereas soybean and common bean are tropical season legumes (Phaseolid clade). Soybean is therefore proposed as a third model legume for both its own economic weight and the phylogenetic proximity to other important crops [4, 5].

Research efforts on model legumes and especially on M. truncatula encompass a broad range of fields in plant biology from population biology [6–8] and plant development [9–16] to plant pathology [17–22], insect resistance [23–27], and biotechnology production [28]. The goal of this review is to provide an overview of the natural characteristics and genetic and genomic tools that make M. truncatula such a desirable experimental system for a growing number of plant biologists. We will highlight how information gained from M. truncatula can be transferred to other legumes crops through comparative genomics and we will share our vision of how M. truncatula can allow us to reach the goal of sustainable well-being through sustainable food and biofuel production.

MEDICAGO TRUNCATULA AS A MODEL LEGUME

Natural attributes of M. truncatula that make it a valuable genetic model include its annual habit and rapid life cycle, its diploid () and autogamous nature, its prolific seed production, and a relatively small genome of about 550 Mb. Jemalong A17 has been selected by the research community as a reference line for most genetic and genomic approaches and is derived from the major commercial cultivar. M. truncatula is native to the Mediterranean basin and is found in a wide range of habitats. It is therefore not surprising to find a high level of variation among and within natural populations [29]. Using microsatellite markers, a publicly available core-collection of 346 inbred lines was developed and thus represents the breadth of this natural diversity [30]. M. truncatula is used as a fodder crop in ley-farming systems in Australia, and a large and diverse collection is housed at the South Australian Research and Development Institute (SARDI) [8].

Like many higher plants, M. truncatula forms symbiotic associations with a wide array of arbuscular mycorrhizal (AM) fungi. As a legume, M. truncatula is also able to develop root nodules with Sinorhizobium melilоti, which is one of the best-characterized rhizobium species at the genetic level [31]. Cultivation-independent techniques have been used to sample the diversity of microbes associated with M. truncatula roots at various developmental stages and they reveal an extremely dynamic genetic structure of its rhizosphere [32].

Mutagenesis approaches using ethyl methane sulfonate (EMS), gamma rays, and fast neutron bombardment (FNB) have generated large mutant populations of M. truncatula from which mutants affected in symbiotic as well as developmental pathways have been identified [33–37]. T-DNA and Tnt1 mutagenesis have been developed recently to generate tagged mutants for forward and reverse genetics purposes [38–40].

Several protocols have been optimized to transform M. truncatula using Agrobacterium tumefaciens [41–45]. These protocols are particularly efficient for the R108 and Jemalong 2HA lines but the regeneration efficiency still needs to be improved for Jemalong A17. This moderate efficiency as well as the time required for the regeneration steps is driving the preference of the Medicago community towards Tnt1 versus T-DNA for gene tagging approaches [46] as well as the search for alternative transformation systems.

Hairy root transformation via Agrobacterium rhizogenes proved to be a rapid and efficient transformation system allowing the generation of transgenic roots in 2-3 weeks. Such "hairy roots" can be infected by rhizobia or AM fungi with symbiotic phenotypes indistinguishable from nontransgenic roots and are therefore an ideal system for plant-microbe symbiosis studies [47]. The development of DsRed as a visual reporter reduced the need for Kanamycin or Basta selection systems which were significantly decreasing nodulation efficiency. This hairy root transformation system is now used routinely to express protein fusions or RNA interference (RNAi) constructs [48–50]. The possibility to regenerate transgenic plants from hairy roots of the R108 line has been reported recently. This flexible approach should allow a rapid initial screening of phenotypes on hairy roots and a subsequent regeneration of transgenic plants if necessary [41]. An interesting ex vitro procedure that eliminates the need for labor-intensive in vitro culture will undoubtedly increase the throughput of hairy root transformations to a level compatible with genomic studies [51]. The Medicago community has therefore identified many ecotypes and developed a wide range of mutants and transgenic lines. A current goal of the International Medicago truncatula steering committee is to address the need for a stock center able to maintain, amplify, and distribute these lines to an ever growing community.

MAPPING THE GENOME OF MEDICAGO TRUNCATULA

Genetic and cytogenetic tools have been instrumental to the development of a "gene rich" genome sequence for M. truncatula. This project also required several bacterial artificial chromosomes (BAC) libraries that were developed using HindIII and EcoRI partial digests as well as a robust physical map (Figure 1).

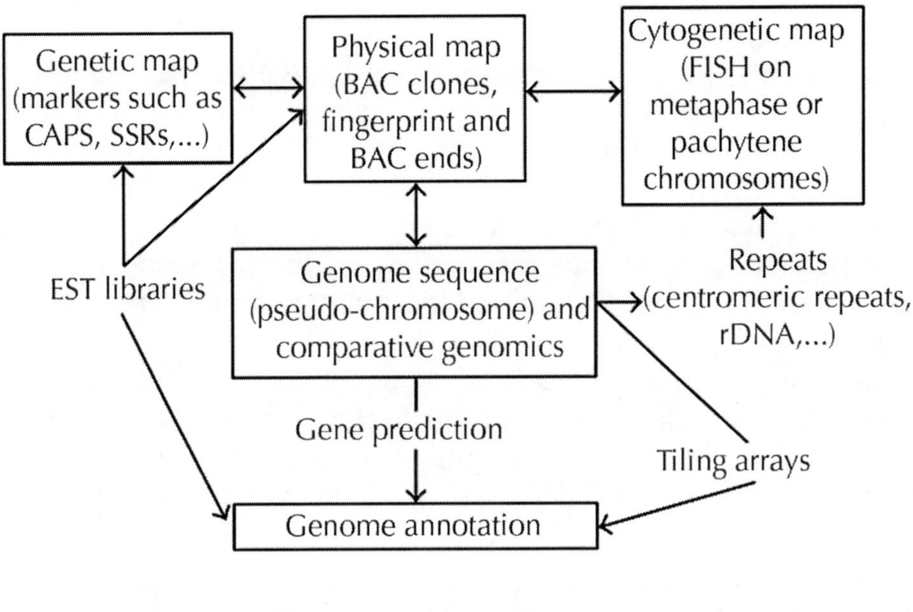

Figure 1: Integration of different maps and libraries to generate and annotate the genomic sequence of M. truncatula. Expressed sequence tags (EST) are used to generate genetic markers and to identify BAC clones in gene-rich regions as well as for gene identification. Repeats identified via genome sequencing and comparison with other species can be mapped via FISH on chromosome spreads.

Genetic maps have been developed from F2 populations and a wide array of genetic markers such as CAPS, AFLPs, RAPDs, and microsatellites (SSRs) [52–54]. One of them, based on a Jemalong A17 A17×A20 F2 population, is currently used as a reference for the genome sequencing project (http://www.medicago.org/genome/map.php). Unfortunately, these F2 populations are either based on a limited amount of genomic DNA or require a labor-intensive vegetative propagation of F2 individuals. In order to provide sustainable tools to the community, genetic maps based on recombinant inbred lines (RILs) and highly polymorphic microsatellite markers are developed and will undoubtedly represent the future reference for M. truncatula genetics (T. Huguet, personal communication).

Cytogenetic maps based on fluorescence in situ hybridization (FISH) with interphase or metaphase chromosomes provide a quick access to the chromosomal location of BAC clones and repeated sequences [55–57]. Obtaining pachytene chromosomes is more labor intensive than metaphase chromosomes but provides an unequalled resolution all along the chromosome and particularly in euchromatic regions [58]. Information from such cytogenetic tools was instrumental for comparative genomics and map-based cloning projects but also allowed the determination that M. truncatula heterochromatin was mostly localized in pericentromeric regions. Genetic and cytogenetic markers corresponding to the borders of these regions have been developed [57, 59, 60]. Based on this unique chromosomal structure, it is therefore possible to predict through the genetic map if a BAC clone belongs to a euchromatic or a heterochromatic region. This observation as well as the possibility to select EST-rich BAC clones led the M. truncatula community to initiate the sequencing of euchromatic (gene-rich) regions via a BAC-by-BAC strategy (http://www.medicago.org/genome/).

Four centers share the sequencing effort of the 8 chromosomes: Bruce Roe et al. at the University of Oklahoma, Chris Town et al. at The Institute for Genomic Research (TIGR), Jane Rogers et al. at the Sanger Centre, and Francis Quétier et al. at the Genoscope. A physical map grouping and ordering of BAC clones was developed by the laboratory of Douglas R. Cook by combining HindIII digestion fingerprints with BAC-end sequence data through the FPC software [52, 61, 62]. More than 1370 FPC contigs cover 466 Mbp (93% of the genome) and are used to determine the minimum tiling path of gene-rich regions for whole genome sequencing [52]. As of February 2007, 188 Mb of genome sequence from 1950 BAC clones are publicly available. About 10% of this information is redundant due to the overlap of BAC clones necessary to create a tiling path and more than 300 gaps between contigs need to be filled. These gaps are sized by FISH and covered with contigs by long-range PCR or classical chromosome walking [62].

Integration of genetic, cytogenetic, physical, and sequence maps allowed the development of pseudochromosomes and greatly facilitated comparative mapping [52, 58–60]. Annotating pseudochromosome sequences is classically achieved through gene prediction programs and comparison with EST databases (Figure 1). The IMGAG (International Medicago Genome Annotation Group) has developed a unique automated pipeline to predict gene structures and functions [63]. More than 25 000 genes have been predicted so far and techniques to test these predictions need to be developed.

Oligonucleotides covering the entire sequence of pseudochromosomes can be printed on glass slides to generate tiling arrays. These arrays can be used for a wide range of applications from gene identification and detection of alternative splicing to comparative genome hybridization (CGH) and chromatin immunoprecipitation on chips (ChIP chips) [64–67].

SYSTEMS ANALYSIS

Transcriptomics

Large-scale EST sequencing is essential for functional genomics studies, permitting the direct identification of large gene collections and setting the stage for further analysis, such as those using DNA microarray technology. Several large EST projects have been completed [68–71]. The analysis of the almost 200 000 ESTs isolated from many different libraries constructed from diverse stages and treatments that came out of these projects is facilitated by searchable databases such as MtDB2 [72] and the TIGR Gene Index (http://www.tigr.org).

Both microarray and macroarray analyses of gene expression changes during symbiosis have been published [73–78]. These experiments ranged from analysis of a few thousand genes on filters during AM symbiosis [73] to almost 10 000 genes compared between wild type and nonnodulating mutants [76, 77] or between fix-mutants [79]. A dual symbiosis chip containing 10 000 M. truncatula genes and the entire S. meliloti prokaryotic genome allows side by side analysis of both partners in the symbiosis [80], and an Affymetrix chip with bioinformatically optimized oligonucleotides representing 48 000 genes is available (http://www.affymetrix.com/support/technical/datasheets/medicago_datasheet.pdf). As genome sequencing continues, following the expression of all M. truncatula genes under varying conditions should soon be possible. Affymetrix placed probe sets for 1850 M. sativa transcripts on these chips to facilitate the study of closely related species such as M. sativa. The use of M. truncatula arrays for analysis of M. sativa (crop alfalfa) gene expression has proven effective [81, 82].

Other effective genomic approaches to transcriptional analysis utilized to date in M. truncatula include suppressive subtractive hybridization (SSH) and serial analysis of gene expression (SAGE). In SSH, suppressive PCR is used to both normalize the abundance of transcripts in individual libraries and enrich for transcripts unique to the library by subtracting sequences common to several libraries, with rare sequences being enriched up to 1000 folds [83]. This method has been used to identify AM specific transcripts [84] and transcripts specifically involved in the S. meliloti symbiosis [85]. SAGE is a method for comprehensive analysis of gene expression patterns using short sequence tags obtained from a unique position within each transcript (10–14 bp) to uniquely identify a transcript. The expression level of the corresponding transcript is determined by quantifying of the number of times a particular tag is observed [86]. Although no publications have arisen yet, a project applying SAGE to M. truncatula is underway at the Center for Medicago Genomics Research at the Nobel Foundation (http://www.noble.org/medicago/GEP.html).

Proteomics

Another complementary approach to identify import gene products involved in interesting processes is to look at changes in the protein complement of a genome that vary by cell or treatment. In order for proteomic approaches to be useful in a system, a large sequence resource is necessary to match the sequences of peptides generated in tryptic digests to their proteins of origin. The growing sequence resource in M. truncatula allows identification of proteins by their mass spectra, making proteomics an effective approach for M. truncatula and proteomics approaches have become quite popular. A comprehensive review of considerations important in proteomics technology and applications in M. truncatula and Arabidopsis was recently published [87, 88]. Because small peptides have been shown to have roles in plant signaling, proteomics has been applied to identifying small protein/peptide components of certain M. truncatula tissues [89]. Proteomic approaches have also been applied to analyses of seed development [14, 16], pathogen interactions [90], symbiosome membranes [91], AM membranes [92], root microsomes [93], and other organ, tissue, and treatment-specific approaches [11, 94–100].

Most of the genes cloned thus far in the initial signal transduction pathway for nodulation are kinases [101]; suggesting global analysis of phosphoproteins is a way to identify important genes involved in signal transduction in M. truncatula. Unfortunately, phosphoproteins involved in cellular signaling are generally present in low abundance, creating new challenges for proteomics. By making adjustments the basic proteomics procedures, such as adding an enrichment step, a proof of concept experiment in M. truncatula phosphoproteomics, gives a taste of the potential of this approach [102].

Metabolomics

Alfalfa produces a number of secondary metabolites of great interest because of their contributions to human health and animal forage quality. The principle behind metabolomics is that metabolic profiling on a genomic scale offers a view of the metabolic status of an organism, which can lend insight to the study gene function or whole plant biology [103]. Successful attempts to link proteomics, transcriptomics, and metabolomics for cell cultures in M. truncatula have emerged from these studies [104, 105].

Metabolomics is a new and evolving science, and requires specialized equipment and multifaceted technical strategies. The Nobel Foundation employs a strategy that utilizes sequential or selective extraction followed by parallel analyses. The parallel analyses achieve a comprehensive view of the metabolome with high-performance liquid chromatography (HPLC), capillary electrophoresis

(CE), gas chromatography (GC), mass spectrometry (MS), and various combinations of the above techniques such as GC/MS, LC/MS, and CE/MS. In addition to studying biological responses to biotic and abiotic elicitors in M. truncatula cellcultures, these techniques are being applied to the study of natural variants in M. truncatula, M. truncatula development, lignin biosynthesis, and legume-insect interactions.

Perhaps the most daunting aspect of metabolomic experiments is the analysis of the data. Early on, it became obvious that metabolomics required a standard similar to MIAME (minimum information about a microarray experiment) to allow comparison of data. A framework for the description of plant metabolomic experiments and their results has recently been developed. ArMet (architecture for metabolomics) is published and in accepted use [106, 107].

Phenomics

As more and more researchers use M. truncatula as a model, the need for a standardized method of describing phenotypes becomes acute. Since the timing and structure of vegetative and floral development in M. truncatula differ from Arabidopsis, adoption of standards such as those used for Arabidopsis [108] is inappropriate. Additionally, M. truncatula symbioses with AM fungi and Sinorhizobium meliloti add another dimension to developmental processes that require a standardized description of process stages and plant anatomy.

To date, a few attempts have been made to develop a standardized language for comparison. Vegetative growth parameters were carefully measured to provide a benchmark in [109], but the use of a glasshouse environment rather than a controlled light and temperature regime rendered the data not universally applicable. Likewise, flower development and response to vernalization have been documented in the same way [109], again in a glasshouse so the light intensity was uncontrolled. These experiments are progress toward a controlled standard for comparison of mutant phenotypes such as "late flowering" or "increased internodal distance." Precision in phenotypic descriptions will be critical to genome scale mutant hunts.

There is no plant structural GO ontology terms for nodulation or nodule structures in the Plant Ontology Consortium site as of the February 2007 release (http://www.plantontology.org). The present plant ontology system provides terms for growth and developmental stages, as well as organs and tissues of Arabidopsis, maize and rice, but none of these plants nodulates, creating a problem for using GO annotation in M. truncatula.

Bioinformatics

All of the "omics" scale tools discussed above necessitate strong bioinformatics infrastructure for the species. A good place to begin is the Medicago Consortium website: http://www.medicago.org. In addition to a handbook of protocols for everything from growing and transforming M. truncatula to naming genes, links from this page lead to informatics tools such as ENSEMBL which allow a real time view of the annotation of the genome, tools allowing browsing of the genome for markers, genes, the location of BACs, the status of the sequencing project or the sequence status of any individual BAC. Users can also view the contigs assembled for sequencing, and make comparisons to other legumes through the legume information system [110] and the consensus legume database (www.legumes.org). Tools are also available through links from the medicago.org website for examining ESTs (TIGR, MtDB2, MENS), and in the future, examining microarray data. In silico approaches in M. truncatula have led to important insights, such as the identification of a large family of small legume-specific transcripts with conserved cysteine motifs whose function continues to be investigated [111–113].

But as genome scale biology is applied, the need to synthesize transcriptomics data, proteomics data, metabolomics data, and more becomes as important as the availability of informatics tools to analyze these data individually. Several steps in this direction have occurred within the M. truncatula community. Some of these integrated solutions are focused around a process, such as gene expression in mycorrhizal symbiosis [114]. Because of the nearly complete genome sequence and the cooperative nature of the sequencing and annotation of the genome, comprehensive integration of various data sources has been necessary from the beginning. Cannon et al. [115] provide a nice summary of the available sequence-based resources and how they interact. A freely available database of biochemical pathway data for M. truncatula (MedicCyc) contains more than 250 pathways with related genes, enzymes, and metabolites [116]. This provides the ability to not only visualize metabolomics data and integrate them with functional genomics data, but also allow comparison of M. truncatula pathways to those in other plants using the compatible AraCyc and RiceCyc databases.

REVERSE GENETICS

Reverse genetics approaches which identify mutants in a gene of interest based on sequence differences are critical genomic tools in a model system. A range of approaches are available, including retrotransposon tagging, T-DNA tagging, TILLING for EMS mutations, PCR screening for fast neutron mutations, and RNA-induced gene silencing (RNAi) [46]. Each method has advantages and disadvantages, and the choice of which method(s) to use will depend on the purpose of the investigator. In M. truncatula, RNAi, TILLING, and PCR screening of Tnt1 insertion mutagenesis populations or fast-neutron generated deletion populations are reverse genetic approaches presently possible.

As noted above, the efficacy of RNAi in M. truncatula has been documented [48] including use in whole plants and in transformed roots. The combination of RNAi constructs and hairy root transformation is useful for large scale screening projects to identify genes of interest for further analysis. A large-scale project to identify gene function by silencing in M. truncatula is underway (http://www.cbs.umn.edu/labs/ganttlab/rnai.html). Initial results from this project include identification of a calcium-dependent protein kinase involved in nodule development, a gene that had not been identified through classical mutational analysis [117].

TILLING (Targeting Induced Local Lesions in Genomes) has proven useful in Arabidopsis and other plants (reviewed in [118]). Briefly, the sequence of a gene of interest is analyzed with a computer program that determines the consequences of all possible EMS mutations (primarily G to A transitions) on the amino acid sequence of the deduced protein. Regions are chosen for PCR amplification based on the concept that those regions most likely to result in altered protein function are highly conserved domains in proteins, and PCR primers are designed to amplify these regions of DNA from each plant. The PCR products are analyzed for single-base pair changes in a high throughput sequencing gel system using an enzyme that detects and cleaves single-base mismatches in DNA. The use of high throughput methodology and a well-characterized and curated population of mutagenized plants allows a plant containing a lesion in the gene of interest to be identified in days. The benefits of TILLING are not only the rapid identification of lesions, but the nature of the lesions themselves. The point mutations generated by EMS treatment allow the use of TILLING to generate an allelic series that includes both missense and nonsense mutations.

In M. truncatula, the Cook lab. at UC Davis developed a population of ~4000 curated EMS mutagenized plants for purpose of TILLING. This resource is currently unavailable as a community resource due to the absence of funding. To date genotypic screens for mutations in 15 genes of interest to the Cook lab or collaborators have been undertaken, and 143 mutants identified, with recovery rates of 9.89 alleles per kbp screened. Phenotypic characterization of one (of 23 unique) allele identified from one of the early genotypic screens for mutations in the M. truncatula arbuscule specific phosphate transporter MtPT4 is described by Javot et al. [119]. Characterization of other mutants in this collection is currently ongoing or advanced to the stage where manuscripts are in preparation for submission (Douglas R. Cook and Varma Penmetsa, personal communication).

A reverse-genetics platform has been established in Medicago truncatula exploiting fast neutron (FN) mutagenesis and a highly sensitive PCR-based detection first documented in Arabidopsis [120]. The FN-based screening platform produces complete loss of function mutants by identifying large deletions in the targeted region. Central to this platform is the development of a detection strategy which allows a mutant amplicon, possessing an internal deletion, to be preferentially amplified in pools where genomic target sequence is present at a 20000-fold excess. This detection sensitivity has been achieved through a combination of techniques for suppressing the amplification of the wild-type sequence and preferentially amplifying the mutant product. The population has been arrayed such that 12000 M2 plants can be analyzed in 4 PCR reactions. These megapools can then be dissected using 25 PCR reactions on 3D pools, allowing identification of the individual seed lot containing the mutant. In comparison with the well-established TILLING method [121–123], which utilizes 8-fold PCR-based screening, FN alleles can be isolated at a fraction of the cost and avoid the problems associated with EMS mutagenesis of targeting small genes and the very high number of background mutations in isolated mutants. An initial characterization of the FN system analyzed 10 genes in a subpopulation of 60 000 M2 plants. Mutants were recovered for 4 target genes. A population of 180 000 M2 plants has now been established and should allow the recovery of mutants from a majority of targeted loci. Information for accessing this resource can be found at www.jicgenomelab.co.uk. (C. Rodgers and G. E. D. Oldroyd, personal communication).

Recently, researchers have identified a tobacco retrotransposon, Tnt1, that moves randomly in M. truncatula but only upon passage through tissue culture [39]. This retrotransposon can be used to generate a large population of plants with tagged mutation sites in tissue culture that become stable upon regeneration of whole plants, an important resource for both forward and reverse genetics. A population mutagenized by Tnt1 can be used for reverse genetic screens by sequencing of tagged sites and forward genetic screens by observation of phenotypes. The isolation of the M. truncatula pim gene through this reverse genetics approach demonstrates the utility of the system for identifying mutants by sequence [40].

TRANSLATIONAL GENOMICS FROM MODEL TO CROP LEGUMES

The value of the model systems will be enhanced by the ability to connect model systems to crops at the structural and functional genome levels. For example, conserved genome structure (synteny) between model and crop species could allow the use of model species as a surrogate genome for map-based cloning of agronomically important genes in crops with complex genomes. Moreover, detailed knowledge of the molecular basis of conserved phenotypes in model species can be translated to great advantage for gene discovery in related species. Working with M. truncatula as a reference system, researchers have used comparative genomics tools to bridge model and crop legumes through comparative mapping of orthologous genes [54, 124, 125]. Alignment of linkage maps and sequenced orthologous regions reveals an extensive network of macro- and microsynteny between legume species [125–127]. In fact, the conserved genome organization between M. truncatula and crop legumes has allowed for cross-species prediction and isolation of several genes required for root symbiosis using M. truncatula as a surrogate [128, 129] and reviewed in [125]. Despite the emerging picture of substantial synteny between legumes, the level of conservation decreases as the evolutionary distance increases [124, 125]. Thus, comparisons within Galegoid or Phaseolid legumes tend to reveal chromosome-level synteny, while comparisons between the two clades tend to reveal large-segment synteny, which is also reflected in the differences in chromosome number between Galegoid and Phaseolid legumes [125, 127].The broad taxonomic distance separating the two clades warrants the development of one or two reference systems within each clade, M. truncatula, and L. japonicus for the cool-season legumes and soybean for the tropical-season legumes [130].

A significant effort has been undertaken in comparative genomic analysis of legume resistance gene homologs (RGHs). Most plant disease resistance genes identified to date belong to the nucleotide binding site (NBS) leucine rich repeat (LRR) family [131]. NBS-LRR genes can be further classified by the presence or absence of a toll/interleukin receptor (TIR) homology domain. In previous studies, researchers investigated the genomic architecture of RGHs in M. truncatula [21], and used phylogenetic methods to assess evolutionary trends in this large gene family in legumes and across the angiosperms [21, 132]. The results from these studies revealed several important insights into RGH gene evolution in plants. Despite the presence of the two major lineages of RGHs (i.e., TIR and non-TIR NBS-LRR genes) in all dicots, each of these lineages is populated by numerous family-specific or family-predominant clades [132]. For example, the major RGH clades that define legumes are absent from the Brassicaceae and Solanaceae, and vice versa. Thus, there are likely to be aspects of RGHs (including both structural and functional attributes) that are peculiar to individual plant families. When phylogenetic analyses were conducted within the legume family [21], it was found that all known major clades in legumes are represented by sequences from M. truncatula, providing evidence that the major RGH radiations predate the respective speciation events. There are also cases that cophyletic RGHs occupy syntenic positions between legumes. The availability of a nearly complete catalog of M. truncatula NBS-LRR genes is expected to greatly enable rapid and efficient characterization of RGHs in other closely related legumes. A legume genome project towards this effort has recently been funded by the NSF Plant Genome Research Program. The goal of this funded project was to develop genomic tools for five less-studied legume species (i.e., chickpea, pigeon pea, cowpea, peanut, and lupine), which are economically important in the developing countries of Africa and Asia (D. R. Cook, personal communication).

Forage legumes, such as alfalfa, red clover, and white clover, are an important component of animal and sustainable agriculture throughout the world. In addition to providing superior forage quality for animal production and improving soil fertility through nitrogen fixation, forage legumes also contribute to the improvement of soil structure and control of soil erosion. Alfalfa (Medicago sativa), for example, is grown on over 26 million acres and ranks third in acreage planted and dollar value in the US (USDA Crop Values Summary 2005). The true clovers (Trifolium spp.), which are often grown together with forage grasses, are also widely distributed. Despite serving as a major source of meat and milk products via animals, the economic importance of forages to food production and the agricultural economy of the US are not fully appreciated. Consequently, forage legumes suffer from poorly developed genetic and genomic infrastructure due to both limited federal funding and their intractable genetic system (e.g., polyploidy and self-incompatibility). The lack of such infrastructure limits the application of genomics-enabled technologies in the genetic improvement of forage legumes. Nevertheless, all these forage legumes are closely related to the model legume M. truncatula, a cool-season legume within the tribe Trifolieae. Therefore, forage legumes could be an immediate beneficiary of the study of M. truncatula genomics. As many of the pathogens of M. truncatula are also pathogens of closely related forage legumes, it should be possible to clone resistance genes that are active against pathogens of crop legume species in M. truncatula. In addition, due to the close relationship of resistance gene sequences between these species, it is likely that functional resistance genes can be moved across species boundaries by transgenic approaches.

Thus the genetic, genomic, and molecular tools available in M. truncatula allow not only investigation of basic processes important to legumes, but also transfer of that information to important crop species.

ACKNOWLEDGMENTS

The authors thank Dr. Kannan Iyyemperumal for his critical reading of the manuscript and useful discussion, and Giles E. D. Oldroyd, Christian Rodgers, Varma Penmetsa, and Douglas R. Cook for sharing unpublished information. Technical contribution no. 5319 of the Clemson University Experiment Station is highly acknowledged.

REFERENCES

1. R. M. Polhill, "Classification of the Leguminosae," in Phytochemical Dictionary of the Leguminosae, F. A. Bisby, J. Buckingham, and J. B. Harborne, Eds., vol. 1, pp. 35–56, Chapman & Hall, New York, NY, USA, 1994.
2. P. H. Graham and C. P. Vance, "Legumes: importance and constraints to greater use," Plant Physiology, vol. 131, no. 3, pp. 872–877, 2003.
3. S. M. de Faria, G. P. Lewis, J. I. Sprent, and J. M. Sutherland, "Occurrence of nodulation in the Leguminosae," New Phytologist, vol. 111, no. 4, pp. 607–619, 1989.
4. S. A. Jackson, D. Rokhsar, G. Stacey, R. C. Shoemaker, J. Schmutz, and J. Grimwood, "Toward a reference sequence of the soybean genome: a multiagency effort," Crop Science, vol. 46, 1, pp. S-55–S-61, 2006.
5. G. Stacey, L. Vodkin, W. A. Parrott, and R. C. Shoemaker, "National Science Foundation-sponsored workshop report. Draft plan for soybean genomics," Plant Physiology, vol. 135, no. 1, pp. 59–70, 2004.
6. Bonnin, J.-M. Prospéri, and I. Olivieri, "Genetic markers and quantitative genetic variation in Medicago truncatula (Leguminosae): a comparative analysis of population structure," Genetics, vol. 143, no. 4, pp. 1795–1805, 1996.
7. Bonnin, J. Ronfort, F. Wozniak, and I. Olivieri, "Spatial effects and rare outcrossing events in Medicago truncatula (Fabaceae)," Molecular Ecology, vol. 10, no. 6, pp. 1371–1383, 2001.

8. S. R. Ellwood, N. K. D'Souza, L. G. Kamphuis, T. I. Burgess, R. M. Nair, and R. P. Oliver, "SSR analysis of the Medicago truncatula SARDI core collection reveals substantial diversity and unusual genotype dispersal throughout the Mediterranean basin," Theoretical and Applied Genetics, vol. 112, no. 5, pp. 977–983, 2006.
9. N. Imin, M. Nizamidin, T. Wu, and B. G. Rolfe, "Factors involved in root formation in Medicago truncatula," Journal of Experimental Botany, vol. 58, no. 3, pp. 439–451, 2007.
10. R. J. Rose, X.-D. Wang, K. E. Nolan, and B. G. Rolfe, "Root meristems in Medicago truncatula tissue culture arise from vascular-derived procambial-like cells in a process regulated by ethylene," Journal of Experimental Botany, vol. 57, no. 10, pp. 2227–2235, 2006. View at Publisher · View at Google Scholar
11. P. Holmes, R. Farquharson, P. J. Hall, and B. G. Rolfe, "Proteomic analysis of root meristems and the effects of acetohydroxyacid synthase-inhibiting herbicides in the root of Medicago truncatula," Journal of Proteome Research, vol. 5, no. 9, pp. 2309–2316, 2006.
12. Q. C. B. Cronk, "Legume flowers bear fruit," Proceedings of the National Academy of Sciences of the United States of America, vol. 103, no. 13, pp. 4801–4802, 2006.
13. B. Bucciarelli, J. Hanan, D. Palmquist, and C. P. Vance, "A standardized method for analysis of Medicago truncatula phenotypic development," Plant Physiology, vol. 142, no. 1, pp. 207–219, 2006.
14. J. Boudet, J. Buitink, F. A. Hoekstra, et al., "Comparative analysis of the heat stable proteome of radicles of Medicago truncatula seeds during germination identifies late embryogenesis abundant proteins associated with desiccation tolerance," Plant Physiology, vol. 140, no. 4, pp. 1418–1436, 2006.
15. N. Djemel, D. Guedon, A. Lechevalier, et al., "Development and composition of the seeds of nine genotypes of the Medicago truncatula species complex," Plant Physiology and Biochemistry, vol. 43, no. 6, pp. 557–566, 2005.
16. K. Gallardo, C. Le Signor, J. Vandekerckhove, R. D. Thompson, and J. Burstin, "Proteomics of Medicago truncatula seed development

establishes the time frame of diverse metabolic processes related to reserve accumulation," Plant Physiology, vol. 133, no. 2, pp. 664–682, 2003.

17. B. Tivoli, A. Baranger, K. Sivasithamparam, and M. J. Barbetti, "Annual Medicago: from a model crop challenged by a spectrum of necrotrophic pathogens to a model plant to explore the nature of disease resistance," Annals of Botany, vol. 98, no. 6, pp. 1117–1128, 2006.
18. F. Colditz, H.-P. Braun, C. Jacquet, K. Niehaus, and F. Krajinski, "Proteomic profiling unravels insights into the molecular background underlying increased Aphanomyces euteiches of Medicago truncatula,"Plant Molecular Biology, vol. 59, no. 3, pp. 387–406, 2005.
19. G. Vandemark and N. Grünwald, "Reaction of Medicago truncatula to Aphanomyces euteiches race 2,"Archives of Phytopathology and Plant Protection, vol. 37, no. 1, pp. 59–67, 2004.
20. O. Nyamsuren, F. Colditz, S. Rosendahl, et al., "Transcriptional profiling of Medicago truncatula roots after infection with Aphanomyces euteiches (oomycota) identifies novel genes upregulated during this pathogenic interaction," Physiological and Molecular Plant Pathology, vol. 63, no. 1, pp. 17–26, 2003.
21. H. Zhu, S. B. Cannon, N. D. Young, and D. R. Cook, "Phylogeny and genomic organization of the TIR and non-TIR NBS-LRR resistance gene family in Medicago truncatula," Molecular Plant-Microbe Interactions, vol. 15, no. 6, pp. 529–539, 2002.
22. F. Vailleau, E. Sartorel, M.-F. Jardinaud, et al., "Characterization of the interaction between the bacterial wilt pathogen Ralstonia solanacearum and the model legume plant Medicago truncatula," Molecular Plant-Microbe Interactions, vol. 20, no. 2, pp. 159–167, 2007.
23. J. P. Klingler, O. R. Edwards, and K. B. Singh, "Independent action and contrasting phenotypes of resistance genes against spotted alfalfa aphid and bluegreen aphid in Medicago truncatula," New Phytologist, vol. 173, no. 3, pp. 630–640, 2007.
24. L.-L. Gao, R. Horbury, R. M. Nair, K. B. Singh, and O. R. Edwards, "Characterization of resistance to multiple aphid species

(Hemiptera: Aphididae) in Medicago truncatula," Bulletin of Entomological Research, vol. 97, no. 1, pp. 41–48, 2007.

25. K. L. Korth, S. J. Doege, S.-H. Park, et al., "Medicago truncatula mutants demonstrate the role of plant calcium oxalate crystals as an effective defense against chewing insects," Plant Physiology, vol. 141, no. 1, pp. 188–195, 2006.
26. J. C. Bede, R. O. Musser, G. W. Felton, and K. L. Korth, "Caterpillar herbivory and salivary enzymes decrease transcript levels of Medicago truncatula genes encoding early enzymes in terpenoid biosynthesis," Plant Molecular Biology, vol. 60, no. 4, pp. 519–531, 2006.
27. L.-L. Gao, J. P. Anderson, J. P. Klingler, R. M. Nair, O. R. Edwards, and K. B. Singh, "Involvement of the octadecanoid pathway in bluegreen aphid resistance in Medicago truncatula," Molecular Plant-Microbe Interactions, vol. 20, no. 1, pp. 82–93, 2007.
28. R. Abranches, S. Marcel, E. Arcalis, F. Altmann, P. Fevereiro, and E. Stoger, "Plants as bioreactors: a comparative study suggests that Medicago truncatula is a promising production system," Journal of Biotechnology, vol. 120, no. 1, pp. 121–134, 2005. Bonnin, T. Huguet, M. Gherardi, J.-M. Prosperi, and I. Olivieri, "High level of polymorphism and spatial structure in a selfing plant species, Medicago truncatula (Leguminosae), shown using RAPD markers," American Journal of Botany, vol. 83, no. 7, pp. 843–855, 1996.
29. J. Ronfort, T. Bataillon, S. Santoni, M. Delalande, J. L. David, and J.-M. Prosperi, "Microsatellite diversity and broad scale geographic structure in a model legume: building a set of nested core collection for studying naturally occurring variation in Medicago truncatula," BMC Plant Biology, vol. 6, 2006.
30. F. Galibert, T. M. Finan, S. R. Long, et al., "The composite genome of the legume symbiontSinorhizobium meliloti," Science, vol. 293, no. 5530, pp. 668–672, 2001.
31. C. Mougel, P. Offre, L. Ranjard, et al., "Dynamic of the genetic structure of bacterial and fungal communities at different developmental stages of Medicago truncatula Gaertn. cv. Jemalong line J5," New Phytologist, vol. 170, no. 1, pp. 165–175, 2006.

32. M. Sagan, D. Morandi, E. Tarenghi, and G. Duc, "Selection of nodulation and mycorrhizal mutants in the model plant Medicago truncatula (Gaertn.) after γ-ray mutagenesis," Plant Science, vol. 111, no. 1, pp. 63–71, 1995.
33. V. Benaben, G. Duc, V. Lefebvre, and T. Huguet, "TE7, an inefficient symbiotic mutant of Medicago truncatula Gaertn. cv Jemalong," Plant Physiology, vol. 107, no. 1, pp. 53–62, 1995.
34. J. M. Ané, G. B. Kiss, B. K. Riely, et al., "Medicago truncatula DMI1 required for bacterial and fungal symbioses in legumes," Science, vol. 303, no. 5662, pp. 1364–1367, 2004.
35. L. J. Bright, Y. Liang, D. M. Mitchell, and J. M. Harris, "The LATD gene of Medicago truncatula is required for both nodule and root development," Molecular Plant-Microbe Interactions, vol. 18, no. 6, pp. 521–532, 2005.
36. R. V. Penmetsa and D. R. Cook, "Production and characterization of diverse developmental mutants of Medicago truncatula," Plant Physiology, vol. 123, no. 4, pp. 1387–1398, 2000.
37. M. Scholte, I. D›Erfurth, S. Rippa, et al., "T-DNA tagging in the model legume Medicago truncatula allows efficient gene discovery," Molecular Breeding, vol. 10, no. 4, pp. 203–215, 2002.
38. D›Erfurth, V. Cosson, A. Eschstruth, H. Lucas, A. Kondorosi, and P. Ratet, "Efficient transposition of the Tnt1 tobacco retrotransposon in the model legume Medicago truncatula," Plant Journal, vol. 34, no. 1, pp. 95–106, 2003.
39. R. Benlloch, I. D›Erfurth, C. Ferrandiz, et al., "Isolation of mtpim proves Tnt1 a useful reverse genetics tool in Medicago truncatula and uncovers new aspects of AP1 functions in legumes," Plant Physiology, vol. 142, no. 3, pp. 972–983, 2006.
40. C. Crane, E. Wright, R. A. Dixon, and Z.-Y. Wang, "Transgenic Medicago truncatula plants obtained from Agrobacterium tumefaciens roots and Agrobacterium rhizogenes hairy roots," Planta, vol. 223, no. 6, pp. 1344–1354, 2006.
41. X. Zhou, M. B. Chandrasekharan, and T. C. Hall, "High rooting frequency and functional analysis of GUS and GFP expression in transgenic Medicago truncatula A17," New Phytologist, vol. 162, no. 3, pp. 813–822, 2004.

42. M. Chabaud, F. de Carvalho-Niebel, and D. G. Barker, "Efficient transformation of Medicago truncatulacv. Jemalong using the hypervirulent Agrobacterium tumefaciens strain AGL1," Plant Cell Reports, vol. 22, no. 1, pp. 46–51, 2003.
43. K. Kamaté, I. D. Rodriguez-Llorente, M. Scholte, et al., "Transformation of floral organs with GFP inMedicago truncatula," Plant Cell Reports, vol. 19, no. 7, pp. 647–653, 2000.
44. T. H. Trinh, P. Ratet, E. Kondorosi, et al., "Rapid and efficient transformation of diploid Medicago truncatula and Medicago sativa ssp. falcata lines improved in somatic embryogenesis," Plant Cell Reports, vol. 17, no. 5, pp. 345–355, 1998.
45. M. Tadege, P. Ratet, and K. S. Mysore, "Insertional mutagenesis: a Swiss Army knife for functional genomics of Medicago truncatula," Trends in Plant Science, vol. 10, no. 5, pp. 229–235, 2005.
46. Boisson-Dernier, M. Chabaud, F. Garcia, G. Bécard, C. Rosenberg, and D. G. Barker, "Agrobacterium rhizogenes roots of Medicago truncatula for the study of nitrogen-fixing and endomycorrhizal symbiotic associations," Molecular Plant-Microbe Interactions, vol. 14, no. 6, pp. 695–700, 2001.
47. E. Limpens, J. Ramos, C. Franken, et al., "RNA interference in Agrobacterium rhizogenes roots ofArabidopsis and Medicago truncatula," Journal of Experimental Botany, vol. 55, no. 399, pp. 983–992, 2004.
48. B. K. Riely, G. Lougnon, J. M. Ané, and D. R. Cook, "The symbiotic ion channel homolog DMI1 is localized in the nuclear membrane of Medicago truncatula roots," Plant Journal, vol. 49, no. 2, pp. 208–216, 2007.
49. X. Huo, E. Schnabel, K. Hughes, and J. Frugoli, "RNAi phenotypes and the localization of a protein::GUS Fusion imply a role for Medicago truncatulaPIN genes in nodulation," Journal of Plant Growth Regulation, vol. 25, no. 2, pp. 156–165, 2006.
50. R. Collier, B. Fuchs, N. Walter, W. K. Lutke, and C. G. Taylor, "Ex vitro composite plants: an inexpensive, rapid method for root biology," Plant Journal, vol. 43, no. 3, pp. 449–457, 2005.
51. J.-H. Mun, D.-J. Kim, H.-K. Choi, et al., "Distribution of microsatellites in the genome of Medicago truncatula: a

resource of genetic markers that integrate genetic and physical maps," Genetics, vol. 172, no. 4, pp. 2541–2555, 2006.

52. P. Thoquet, M. Ghérardi, E.-P. Journet, et al., "The molecular genetic linkage map of the model legumeMedicago truncatula: an essential tool for comparative legume genomics and the isolation of agronomically important genes," BMC Plant Biology, vol. 2, 2002.
53. H.-K. Choi, D. Kim, T. Uhm, et al., "A sequence-based genetic map of Medicago truncatula and comparison of marker colinearity with M. sativa," Genetics, vol. 166, no. 3, pp. 1463–1502, 2004.
54. M. Abirached-Darmency, E. Prado-Vivant, L. Chelysheva, and T. Pouthier, "Variation in rDNA locus number and position among legume species and detection of 2 linked rDNA loci in the model Medicago truncatula by FISH," Genome, vol. 48, no. 3, pp. 556–561, 2005.
55. M. Cerbah, Z. Kevei, S. Siljak-Yakovlev, E. Kondorosi, A. Kondorosi, and T. H. Trinh, "FISH chromosome mapping allowing karyotype analysis in Medicago truncatula lines Jemalong J5 and R-108-1," Molecular Plant-Microbe Interactions, vol. 12, no. 11, pp. 947–950, 1999.
56. J. M. Ané, J. Lévy, P. Thoquet, et al., "Genetic and cytogenetic mapping of DMI1, DMI2, and DMI3 genes of Medicago truncatula involved in Nod factor transduction, nodulation, and mycorrhization," Molecular Plant-Microbe Interactions, vol. 15, no. 11, pp. 1108–1118, 2002.
57. O. Kulikova, G. Gualtieri, R. Geurts, et al., "Integration of the FISH pachytene and genetic maps ofMedicago truncatula," Plant Journal, vol. 27, no. 1, pp. 49–58, 2001.
58. O. Kulikova, R. Geurts, M. Lamine, et al., "Satellite repeats in the functional centromere and pericentromeric heterochromatin of Medicago truncatula," Chromosoma, vol. 113, no. 6, pp. 276–283, 2004.
59. E. Schnabel, O. Kulikova, R. V. Penmetsa, T. Bisseling, D. R. Cook, and J. Frugoli, "An integrated physical, genetic and cytogenetic map around the sunn locus of Medicago truncatula," Genome, vol. 46, no. 4, pp. 665–672, 2003.

60. F. W. Engler, J. Hatfield, W. Nelson, and C. A. Soderlund, "Locating sequence on FPC maps and selecting a minimal tiling path," Genome Research, vol. 13, no. 9, pp. 2152–2163, 2003.
61. N. D. Young, S. B. Cannon, S. Sato, et al., "Sequencing the genespaces of Medicago truncatula and Lotus japonicus," Plant Physiology, vol. 137, no. 4, pp. 1174–1181, 2005.
62. C. D. Town, "Annotating the genome of Medicago truncatula," Current Opinion in Plant Biology, vol. 9, no. 2, pp. 122–127, 2006.
63. V. Stolc, M. P. Samanta, and W. Tongprasit, "Identification of transcribed sequences in Arabidopsis thaliana by using high-resolution genone tiling arrays," Proceedings of the National Academy of Sciences of the United States of America, vol. 102, no. 12, pp. 4453–4458, 2005.
64. J. M. Johnson, S. Edwards, D. Shoemaker, and E. E. Schadt, "Dark matter in the genome: evidence of widespread transcription detected by microarray tiling experiments," Trends in Genetics, vol. 21, no. 2, pp. 93–102, 2005.
65. K. Yamada, J. Lim, J. H. Dale, et al., "Empirical analysis of transcriptional activity in the Arabidopsis Genome," Science, vol. 302, no. 5646, pp. 842–846, 2003.
66. P. Bertone, M. Gerstein, and M. Snyder, "Applications of DNA tiling arrays to experimental genome annotation and regulatory pathway discovery," Chromosome Research, vol. 13, no. 3, pp. 259–274, 2005.
67. P. A. Covitz, L. S. Smith, and S. R. Long, "Expressed sequence tags from a root-hair-enriched Medicago truncatula cDNA library," Plant Physiology, vol. 117, no. 4, pp. 1325–1332, 1998.
68. J. Györgyey, D. Vaubert, J. I. Jiménez-Zurdo, et al., "Analysis of Medicago truncatula nodule expressed sequence tags," Molecular Plant-Microbe Interactions, vol. 13, no. 1, pp. 62–71, 2000.
69. C. J. Bell, R. A. Dixon, A. D. Farmer, et al., "The Medicago genome initiative: a model legume database,"Nucleic Acids Research, vol. 29, no. 1, pp. 114–117, 2001.
70. E.-P. Journet, D. van Tuinen, J. Gouzy, et al., "Exploring root symbiotic programs in the model legumeMedicago

truncatula using EST analysis," Nucleic Acids Research, vol. 30, no. 24, pp. 5579–5592, 2002.

71. A.-F. J. Lamblin, J. A. Crow, J. E. Johnson, et al., "MtDB: a database for personalized data mining of the model legume Medicago truncatula transcriptome," Nucleic Acids Research, vol. 31, no. 1, pp. 196–201, 2003.
72. J. Liu, L. A. Blaylock, G. Endre, et al., "Transcript profiling coupled with spatial expression analyses reveals genes involved in distinct developmental stages of an arbuscular mycorrhizal symbiosis," Plant Cell, vol. 15, no. 9, pp. 2106–2123, 2003.
73. F. E. Yahyaoui, H. Küster, B. B. Amor, et al., "Expression profiling in Medicago truncatula identifies more than 750 genes differentially expressed during nodulation, including many potential regulators of the symbiotic program," Plant Physiology, vol. 136, no. 2, pp. 3159–3176, 2004.
74. K. Manthey, F. Krajinski, N. Hohnjec, et al., "Transcriptome profiling in root nodules and arbuscular mycorrhiza identifies a collection of novel genes induced during Medicago truncatula root endosymbioses," Molecular Plant-Microbe Interactions, vol. 17, no. 10, pp. 1063–1077, 2004.
75. R. M. Mitra and S. R. Long, "Plant and bacterial symbiotic mutants define three transcriptionally distinct stages in the development of the Medicago truncatula/Sinorhizobium meliloti Symbiosis," Plant Physiology, vol. 134, no. 2, pp. 595–604, 2004.
76. R. M. Mitra, S. L. Shaw, and S. R. Long, "Six nonnodulating plant mutants defective for Nod factor-induced transcriptional changes associated with the legume-rhizobia symbiosis," Proceedings of the National Academy of Sciences of the United States of America, vol. 101, no. 27, pp. 10217–10222, 2004.
77. D. P. Lohar, N. Sharopova, G. Endre, et al., "Transcript analysis of early nodulation events in Medicago truncatula," Plant Physiology, vol. 140, no. 1, pp. 221–234, 2006.
78. C. G. Starker, A. L. Parra-Colmenares, L. Smith, R. M. Mitra, and S. R. Long, "Nitrogen fixation mutants of Medicago truncatula fail to support plant and bacterial symbiotic gene expression," Plant Physiology, vol. 140, no. 2, pp. 671–680, 2006.

79. M. J. Barnett, C. J. Toman, R. F. Fisher, and S. R. Long, "A dual-genome Symbiosis Chip for coordinate study of signal exchange and development in a prokaryote-host interaction," Proceedings of the National Academy of Sciences of the United States of America, vol. 101, no. 47, pp. 16636–16641, 2004.
80. N. Aziz, N. L. Paiva, G. D. May, and R. A. Dixon, "Transcriptome analysis of alfalfa glandular trichomes," Planta, vol. 221, no. 1, pp. 28–38, 2005.
81. B. E. Deavours and R. A. Dixon, "Metabolic engineering of isoflavonoid biosynthesis in alfalfa," Plant Physiology, vol. 138, no. 4, pp. 2245–2259, 2005.
82. L. Diatchenko, Y.-F. C. Lau, A. P. Campbell, et al., "Suppression subtractive hybridization: a method for generating differentially regulated or tissue-specific cDNA probes and libraries," Proceedings of the National Academy of Sciences of the United States of America, vol. 93, no. 12, pp. 6025–6030, 1996.
83. Wulf, K. Manthey, J. Doll, et al., "Transcriptional changes in response to arbuscular mycorrhiza development in the model plant Medicago truncatula," Molecular Plant-Microbe Interactions, vol. 16, no. 4, pp. 306–314, 2003.
84. L. Godiard, A. Niebel, F. Micheli, J. Gouzy, T. Ott, and P. Gamas, "Identification of new potential regulators of the Medicago truncatula-Sinorhizobium meliloti symbiosis using a large-scale suppression subtractive hybridization approach," Molecular Plant-Microbe Interactions, vol. 20, no. 3, pp. 321–332, 2007.
85. V. E. Velculescu, L. Zhang, B. Vogelstein, and K. W. Kinzler, "Serial analysis of gene expression," Science, vol. 270, no. 5235, pp. 484–487, 1995.
86. G. K. Agrawal, M. Yonekura, Y. Iwahashi, H. Iwahashi, and R. Rakwal, "System, trends and perspectives of proteomics in dicot plants—part III: unraveling the proteomes influenced by the environment, and at the levels of function and genetic relationships," Journal of Chromatography B, vol. 815, no. 1-2, pp. 137–145, 2005.
87. G. K. Agrawal, M. Yonekura, Y. Iwahashi, H. Iwahashi, and R. Rakwal, "System, trends and perspectives of proteomics in dicot plants—part II: proteomes of the complex developmental

stages," Journal of Chromatography B, vol. 815, no. 1-2, pp. 125–136, 2005.

88. K. Zhang, C. McKinlay, C. H. Hocart, and M. A. Djordjevic, "The Medicago truncatula small protein proteome and peptidome," Journal of Proteome Research, vol. 5, no. 12, pp. 3355–3367, 2006.

89. F. Colditz, O. Nyamsuren, K. Niehaus, H. Eubel, H.-P. Braun, and F. Krajinski, "Proteomic approach: identification of Medicago truncatula proteins induced in roots after infection with the pathogenic oomycete Aphanomyces euteiches," Plant Molecular Biology, vol. 55, no. 1, pp. 109–120, 2004.

90. C. M. Catalano, W. S. Lane, and D. J. Sherrier, "Biochemical characterization of symbiosome membrane proteins from Medicago truncatula root nodules," Electrophoresis, vol. 25, no. 3, pp. 519–531, 2004.

91. B. Valot, M. Dieu, G. Recorbet, M. Raes, S. Gianinazzi, and E. Dumas-Gaudot, "Identification of membrane-associated proteins regulated by the arbuscular mycorrhizal symbiosis," Plant Molecular Biology, vol. 59, no. 4, pp. 565–580, 2005.

92. B. Valot, S. Gianinazzi, and D.-G. Eliane, "Sub-cellular proteomic analysis of a Medicago truncatula root microsomal fraction," Phytochemistry, vol. 65, no. 12, pp. 1721–1732, 2004.

93. U. Mathesius, G. Keijzers, S. H. A. Natera, J. J. Weinman, M. A. Djordjevic, and B. G. Rolfe, "Establishment of a root proteome reference map for the model legume Medicago truncatula using the expressed sequence tag database for peptide mass fingerprinting," Proteomics, vol. 1, no. 11, pp. 1424–1440, 2001.

94. B. S. Watson, V. S. Asirvatham, L. Wang, and L. W. Sumner, "Mapping the proteome of barrel medic (Medicago truncatula)," Plant Physiology, vol. 131, no. 3, pp. 1104–1123, 2003. N. Imin, F. De Jong, U. Mathesius, et al., "Proteome reference maps of Medicago truncatula embryogenic cell cultures generated from single protoplasts," Proteomics, vol. 4, no. 7, pp. 1883–1896, 2004.

95. N. Imin, M. Nizamidin, D. Daniher, K. E. Nolan, R. J. Rose, and B. G. Rolfe, "Proteomic analysis of somatic embryogenesis in Medicago truncatula. Explant cultures grown under 6-benzylaminopurine

and 1-naphthaleneacetic acid treatments," Plant Physiology, vol. 137, no. 4, pp. 1250–1260, 2005.

96. Z. Lei, A. M. Elmer, B. S. Watson, R. A. Dixon, P. J. Mendes, and L. W. Sumner, "A two-dimensional electrophoresis proteomic reference map and systematic identification of 1367 proteins from a cell suspension culture of the model legume Medicago truncatula," Molecular and Cellular Proteomics, vol. 4, no. 11, pp. 1812–1825, 2005.
97. N. Amiour, G. Recorbet, F. Robert, S. Gianinazzi, and E. Dumas-Gaudot, "Mutations in DMI3 andSUNN modify the appressorium-responsive root proteome in arbuscular mycorrhiza," Molecular Plant-Microbe Interactions, vol. 19, no. 9, pp. 988–997, 2006.
98. J. Prayitno, B. G. Rolfe, and U. Mathesius, "The ethylene-insensitive sickle mutant of Medicago truncatulashows altered auxin transport regulation during nodulation," Plant Physiology, vol. 142, no. 1, pp. 168–180, 2006.
99. M. K. Udvardi and W.-R. Scheible, "GRAS genes and the symbiotic green revolution," Science, vol. 308, no. 5729, pp. 1749–1750, 2005.
100. S. Laugesen, E. Messinese, S. Hem, et al., "Phosphoproteins analysis in plants: a proteomic approach,"Phytochemistry, vol. 67, no. 20, pp. 2208–2214, 2006.
101. R. N. Trethewey, A. J. Krotzky, and L. Willmitzer, "Metabolic profiling: a rosetta stone for genomics?"Current Opinion in Plant Biology, vol. 2, no. 2, pp. 83–85, 1999.
102. C. D. Broeckling, D. V. Huhman, M. A. Farag, et al., "Metabolic profiling of Medicago truncatula cell cultures reveals the effects of biotic and abiotic elicitors on metabolism," Journal of Experimental Botany, vol. 56, no. 410, pp. 323–336, 2005.
103. H. Suzuki, M. S.S. Reddy, M. Naoumkina, et al., "Methyl jasmonate and yeast elicitor induce differential transcriptional and metabolic re-programming in cell suspension cultures of the model legume Medicago truncatula," Planta, vol. 220, no. 5, pp. 696–707, 2005.
104. H. Jenkins, N. Hardy, M. Beckmann, et al., "A proposed framework for the description of plant metabolomics experiments and their

results," Nature Biotechnology, vol. 22, no. 12, pp. 1601–1606, 2004.

105. H. Jenkins, H. Johnson, B. Kular, T. Wang, and N. Hardy, "Toward supportive data collection tools for plant metabolomics," Plant Physiology, vol. 138, no. 1, pp. 67–77, 2005.
106. J. Bowman, Arabidopsis: An Atlas of Morphology and Development, Springer, New York, NY, USA, 1994.
107. D. Moreau, C. Salon, and N. Munier-Jolain, "A framework for the analysis of genetic variability inMedicago truncatula," in Proceedings of the 2nd International Conference on Legume Genomics and Genetics, p. 286, Dijon, France, June 2004.
108. M. D. Gonzales, E. Archuleta, A. Farmer, et al., "The legume information system (LIS): an integrated information resource for comparative legume biology," Nucleic Acids Research, vol. 33, pp. D660–D665, 2005.
109. M. Fedorova, J. van de Mortel, P. A. Matsumoto, et al., "Genome-wide identification of nodule-specific transcripts in the model legume Medicago truncatula," Plant Physiology, vol. 130, no. 2, pp. 519–537, 2002.
110. P. Mergaert, K. Nikovics, Z. Kelemen, et al., "A novel family in Medicago truncatula consisting of more than 300 nodule-specific genes coding for small, secreted polypeptides with conserved cysteine motifs,"Plant Physiology, vol. 132, no. 1, pp. 161–173, 2003.
111. M. A. Graham, K. A.T. Silverstein, S. B. Cannon, and K. A. VandenBosch, "Computational identification and characterization of novel genes from legumes," Plant Physiology, vol. 135, no. 3, pp. 1179–1197, 2004.
112. H. Küster, A. Becker, C. Firnhaber, et al., "Development of bioinformatic tools to support EST-sequencing, in silico and microarray-based transcriptome profiling in mycorrhizal symbioses,"Phytochemistry, vol. 68, no. 1, pp. 19–32, 2007.
113. S. B. Cannon, J. A. Crow, M. L. Heuer, et al., "Databases and information integration for the Medicago truncatula genome and transcriptome," Plant Physiology, vol. 138, no. 1, pp. 38–46, 2005.

114. E. Urbanczyk-Wochniak and L. W. Sumner, "MedicCyc: a biochemical pathway database for Medicago truncatula," Bioinformatics, vol. 23, no. 11, pp. 1418–1423, 2007
115. J. Liu, S. S. Miller, M. Graham, et al., "Recruitment of novel calcium-binding proteins for root nodule symbiosis in Medicago truncatula," Plant Physiology, vol. 141, no. 1, pp. 167–177, 2006.
116. S. Henikoff, T. Furuyama, and K. Ahmad, "Histone variants, nucleosome assembly and epigenetic inheritance," Trends in Genetics, vol. 20, no. 7, pp. 320–326, 2004.
117. H. Javot, R.V. Penmetsa, N. Terzaghi, D. R. Cook, and M. J. Harrison, "A Medicago truncatulaphosphate transporter indispensable for the arbuscular mycorrhizal symbiosis," Proceedings of the National Academy of Sciences of the United States of America, vol. 104, no. 5, pp. 1720–1725, 2007.
118. X. Li, Y. Song, K. Century, et al., "A fast neutron deletion mutagenesis-based reverse genetics system for plants," Plant Journal, vol. 27, no. 3, pp. 235–242, 2001.
119. T. Colbert, B. J. Till, R. Tompa, et al., "High-throughput screening for induced point mutations," Plant Physiology, vol. 126, no. 2, pp. 480–484, 2001.
120. C. M. McCallum, L. Comai, E. A. Greene, and S. Henikoff, "Targeting induced local lesions IN genomes (TILLING) for plant functional genomics," Plant Physiology, vol. 123, no. 2, pp. 439–442, 2000.
121. C. M. McCallum, L. Comai, E. A. Greene, and S. Henikoff, "Targeted screening for induced mutations,"Nature Biotechnology, vol. 18, no. 4, pp. 455–457, 2000.
122. H.-K. Choi, J.-H. Mun, D.-J. Kim, et al., "Estimating genome conservation between crop and model legume species," Proceedings of the National Academy of Sciences of the United States of America, vol. 101, no. 43, pp. 15289–15294, 2004.
123. H. Zhu, H.-K. Choi, D. R. Cook, and R. C. Shoemaker, "Bridging model and crop legumes through comparative genomics," Plant Physiology, vol. 137, no. 4, pp. 1189–1196, 2005.
124. H. Zhu, B. K. Riely, N. J. Burns, and J. M. Ané, "Tracing

nonlegume orthologs of legume genes required for nodulation and arbuscular mycorrhizal symbioses," Genetics, vol. 172, no. 4, pp. 2491–2499, 2006.

125. S. B. Cannon, L. Sterck, S. Rombauts, et al., "Legume genome evolution viewed through the Medicago truncatula and Lotus japonicus genomes," Proceedings of the National Academy of Sciences of the United States of America, vol. 103, no. 40, pp. 14959–14964, 2006.
126. G. Endre, P. Kaló, Z. Kevei, et al., "Genetic mapping of the non-nodulation phenotype of the mutant MN-1008 in tetraploid alfalfa (Medicago sativa)," Molecular Genetics and Genomics, vol. 266, no. 6, pp. 1012–1019, 2001.
127. E. Limpens, C. Franken, P. Smit, J. Willemse, T. Bisseling, and R. Geurts, "LysM domain receptor kinases regulating rhizobial nod factor-induced infection," Science, vol. 302, no. 5645, pp. 630–633, 2003.
128. P. Gepts, W. D. Beavis, E. C. Brummer, et al., "Legumes as a model plant family. Genomics for food and feed report of the cross-legume advances through genomics conference," Plant Physiology, vol. 137, no. 4, pp. 1228–1235, 2005.
129. S. H. Hulbert, C. A. Webb, S. M. Smith, and Q. Sun, "Resistance gene complexes: evolution and utilization," Annual Review of Phytopathology, vol. 39, pp. 285–312, 2001.
130. S. B. Cannon, H. Zhu, A. M. Baumgarten, et al., "Diversity, distribution, and ancient taxonomic relationships within the TIR and non-TIR NBS-LRR resistance gene subfamilies," Journal of Molecular Evolution, vol. 54, no. 4, pp. 548–562, 2002.

Citations

CHAPTER 1

Miguel Mourato, Rafaela Reis and Luisa Louro Martins (2012). Characterization of Plant Antioxidative System in Response to Abiotic Stresses: A Focus on Heavy Metal Toxicity, Advances in Selected Plant Physiology Aspects, Dr. Giuseppe Montanaro (Ed.), ISBN: 978-953-51-0557-2, InTech, DOI: 10.5772/34557.

CHAPTER 2

Anna M. De Leonardis, Maria Petrarulo, Pasquale De Vita and Anna M. Mastrangelo (2012). Genetic and Molecular Aspects of Plant Response to Drought in Annual Crop Species, Advances in Selected Plant Physiology Aspects, Dr. Giuseppe Montanaro (Ed.), ISBN: 978-953-51-0557-2.

CHAPTER 3

Susanne Berger, Alok K. Sinha, and Thomas Roitsch, Plant physiology meets phytopathology: plant primary metabolism and plant–pathogen interactions, J. Exp. Bot. (2007) 58 (15-16): 4019-4026. doi: 10.1093/jxb/erm298.

CHAPTER 4

J. G ILES WA INES* and BAHMAN EHDA IE, Domestication and Crop Physiology: Roots of Green-Revolution Wheat, doi:10.1093/aob/mcm180.

CHAPTER 5

Giuseppe Montanaro, Bartolomeo Dichio and Cristos Xiloyannis (2012). Fruit Transpiration: Mechanisms and Significance for Fruit Nutrition and Growth, Advances in Selected Plant Physiology Aspects, Dr. Giuseppe Montanaro (Ed.), ISBN: 978-953-51-0557-2, InTech, DOI: 10.5772/21411.

CHAPTER 6

Shilpi Khare Saikia and Rakesh Pandey, "Rhizospheric Microflora Escalating Aroma Constituents and Yield Attributes in Ocimum tenuiflorum (L.) cv. CIM-Ayu," Advances in Agriculture, vol. 2014, Article ID 621912, 7 pages, 2014. doi:10.1155/2014/621912

CHAPTER 7

Huilan Yi, Xin Liu, Min Yi, and Gang Chen, Dual Role of Hydrogen Peroxide in Arabidopsis Guard Cells in Response to Sulfur Dioxide, Advances in Toxicology, vol. 2014, Article ID 407368, 9 pages, 2014. doi:10.1155/2014/407368

CHAPTER 8

Rosa M. Pérez-Clemente, Vicente Vives, Sara I. Zandalinas, María F. López-Climent, Valeria Muñoz, and Aurelio Gómez-Cadenas, "Biotechnological Approaches to Study Plant Responses to Stress," BioMed Research International, vol. 2013, Article ID 654120, 10 pages, 2013. doi:10.1155/2013/654120

CHAPTER 9

Philippe Jeandet, Bertrand Delaunois, Aziz Aziz, et al., "Metabolic Engineering of Yeast and Plants for the Production of the Biologically Active Hydroxystilbene, Resveratrol," Journal of Biomedicine and Biotechnology, vol. 2012, Article ID 579089, 14 pages, 2012. doi:10.1155/2012/579089

CHAPTER 10

Jean-Michel Ané, Hongyan Zhu, and Julia Frugoli, "Recent Advances in Medicago truncatula Genomics," International Journal of Plant Genomics, vol. 2008, Article ID 256597, 11 pages, 2008. doi:10.1155/2008/256597

Index

S

T

U

V

W

Z